Bernhard Irrgang: Critics of Technological Lifeworld

AF061071

Dresden Philosophy of Technology Studies
Dresdner Studien zur Philosophie der Technologie

Edited by/Herausgegeben von Bernhard Irrgang

Vol./Bd. 4

Frankfurt am Main · Berlin · Bern · Bruxelles · New York · Oxford · Wien

Arun Kumar Tripathi (ed.)

Bernhard Irrgang: Critics of Technological Lifeworld

Collection of Philosophical Essays

PETER LANG
Internationaler Verlag der Wissenschaften

Bibliographic Information published by the Deutsche Nationalbibliothek
The Deutsche Nationalbibliothek lists this publication in the Deutsche Nationalbibliografie; detailed bibliographic data is available in the internet at http://dnb.d-nb.de.

Cover Design:
Olaf Gloeckler, Atelier Platen, Friedberg

Photo on Cover:
Kabelsalat in Thailand
Photographer: Steffen Steinert

The publication of this anthology is carried out
with the friendly assistance
of the Department of Philosophy
at the Technische Universität Dresden,
Germany.

ISSN 1861-423X
ISBN 978-3-631-58570-2
© Peter Lang GmbH
Internationaler Verlag der Wissenschaften
Frankfurt am Main 2011
All rights reserved.

All parts of this publication are protected by copyright. Any utilisation outside the strict limits of the copyright law, without the permission of the publisher, is forbidden and liable to prosecution. This applies in particular to reproductions, translations, microfilming, and storage and processing in electronic retrieval systems.

www.peterlang.de

Preface

Today, more than ever before, there is an urgent need to understand the imperative of modernization and its attendant idiom of globalization. We require an understanding of science and technology on the basis of culture, wisdom, ecology and ethical values. The process of current globalization is emerging into a cultural, historical, and ecological phenomenon. At the same time, this change is adding an ethical dimension to the development of technology, which requires a deeper understanding of techniques, technology and science.

The philosophy of technology remains a relatively young sub-discipline in philosophy. Although there were late 19th century intimations with Karl Marx and Ernst Kapp, its origins are largely 20th century. Even the term, „technology," it is a largely 20th century term and use. David Nye, a historian of technology notes that there are very few references to technology in the late 19th century, with „inventions" and „applied arts" being more common until after the First World War And, it is just after the first World War that philosophers, particularly in Europe, began to use the term, „technology," and to write book-length works about or including discussions of technology, Don Ihde argues.

Technologies were analogized as extensions and magnifications of human organic processes and projected into an external environment. A philosophy of technological culture should take the material culture of technology into account. Classical phenomenological philosophy of technology has mainly tried to understand technology in terms of the conditions of its possibility. As Peter-Paul Verbeek (in his book What Things Do) has claimed that, the transcendentalist approach resulted in a one-sided and inadequate understanding of technology in terms of alienation. What is needed is not so much an analysis of the origins of technologies but of what they actually do: the ways, in which they co-shape the relationships between humans and their world, Verbeek argues.

Since 1993, Bernhard Irrgang who is a Professor for the Philosophy of Technology, has been teaching courses in philosophy of technology, medical ethics and ethical hermeneutics at the Institute for Philosophy at Dresden University of Technology (TU Dresden) Germany. The main topics of educational activities of the institute are: Philosophical questions and topics concerning technique and technology; Technology transfer; Technological development and early forms of technological cultures; Technoscience; Science, Technology and Society studies, especially focusing on genetic engineering, cultural theory of technology, Information Technology, Artificial Intelligence and Expert Systems; Technology Assessment; Philosophy and Biology; Hermeneutical Ethics; Technological and Ecological ethics; Medical Ethics; Intercultural environmental ethics; and History of Philosophy in the 17th, 18th, and 20th century.

Irrgang for last 2 decades engaging with the questions, what role does technology play in everyday human experience? How do technological artefacts affect people's existence and their relations with the world? And how do instruments, devices and apparatuses produce and transform human knowledge? These are the central questions in Bernhard Irrgang's philosophy of technology. For the actual discussion on ethics and technologies, Irrgang in his book „Hermeneutics Ethics: Pragmatically and ethical orientation in technological societies" is proposing a philosophical and pragmatic approach of „hermeneutics of lifeworld." Irrgang tries to connect the traditions of hermeneutics and the moral philosophy, which makes the ethical problems of contemporary world (which is technologically mediated) intelligibly understandable. Irrgang is offering an introductory approach for ethical judgments, which facilitates and motivates the human beings to overcome their complex everyday lifeworld and understand technically embedded everyday situations to act responsibly. At the same time, for technological world, Irrgang has also left the new possibilities of understanding to master the everyday lifeworld, whereas human beings need competence.

Technology consists of the constellations of materials, tools, and procedures we use to do things. It did not come into being only in modern times, but has existed as long as man has been on the earth. Indeed, it may be considered an essential aspect of being human. Pre-eminent historian of technology, Carl Mitcham has introduced the basic philosophical issues involved and reveals the scope of the problems of technology as going to the foundations of everyday life and our relations to the world. Along with Don Ihde and Carl Mitcham, Bernhard Irrgang provides a useful vocabulary for understanding the ways we relate to technology and to the world through technology.

It is long argued that, the brute fact of technology in modern society is urging philosophers of technology out from under the shadow of philosophers of science. Along with Don Ihde and Albert Borgmann, Bernhard Irrgang in his works, has responded to the „given fact" that we live in a „technological and organizational culture" by sketching a „praxis philosophy" of technologies. Don Ihde and Bernhard Irrgang, in particular draws his inspiration from Heidegger's „tool analysis", Husserl's notion of „intentionality" and Merleau-Ponty's notion of „lived body." Praxis philosophies (existentialism, phenomenology, Marxism dialectical traditions, and some forms of pragmatism) are, according to Ihde and Irrgang, based on a theory which affirms the importance of „perception and embodiment."

The anthology „Bernhard Irrgang: Critics of Technological Lifeworld: Collection of Philosophical Essays" didn't appear without the intensive support and philosophical guidance of Irrgang and his penchant mentoring on the philosophy of technology. Editor wants to put Irrgang's philosophy of technology between European and American pragmatism.

Bernhard Irrgang, who is a proponent and interlocutor of the diversity of intellectual cultures, is widely acknowledged as one of the most important German phenomenologist philosopher, have been working in the tradition of Heidegger's hermeneutics and Ihde's phenomenology.

Editor would put Irrgang in the tradition of Euro-American phenomenology and hermeneutics. Along with long „Paradigmatic Shifts in the Contemporary philosophy of technologies" this anthology contains 8 essays by Bernhard Irrgang on „Social and Ethical Aspects of Biotechnological Practice", „Technological Development and Social Progress", „Technology transfer as a transcultural modernization: What Can Philosophers of Technology Contribute?", „Ethical Action in Robotics", „Epistemology of Biotechnology", „Justified Trust in Technology", „Postphenomenological investigations to brain research and human-embodied mind", and „Visions of Technology." The anthology has an epilogue „Questions Concerning Technology" by upcoming Dutch philosopher of technology, Peter-Paul Verbeek.

Editor Arun Tripathi, who himself has been working with Bernhard Irrgang for last 8 years sincerely thankful for all the support he received from him. If Bernhard Irrgang didn't exist, then this editor has to discover him in the tradition of philosophy of body and technology.

Arun Kumar Tripathi
TU Dresden, November 10th, 2010

Acknowledgments

Editor Arun Kumar Tripathi and Prof. Bernhard Irrgang want to express their sincere gratitude to Ms. Claire Shadwell, University of Gloucestershire for her help in reading and correcting the translations and making amendments wherever necessary. Editor is very grateful to Ms. Martina Polster of Peter Lang Verlag for her patience and timely assistance.

Editor also wants to express his deep indebtedness to Mr. Michael Funk (BA), Research Assistant (Department of Philosophy of Technology, Institute for Philosophy, Dresden University of Technology, Germany) for his sincere help and assistance in formatting the texts.

Another version of the Chapter 2 „Social and Ethical Aspects of Biotechnological Practice" is published in the ACM Ubiquity. Part of the Chapter 3 „Technological Development and Social Progress" is published in the Seminarios de Filosofia, pp. 1-52, Vol.12-13, 1999-2000.

Another version of the Chapter 5 „Ethical Action in Robotics" is published in the Conference Proceedings of the Sixth International Conference of Computer Ethics: Philosophical Inquiry „Ethics of New Information Technology" (CEPE2005, July 17-19, 2005, Enschede, TheNetherlands) eds. Philip Brey, Frances Grodzinsky and Lucas Introna, CEPTES, Enschede, pp. 241-250. (pp. 241-250).

Another version of the Chapter 6 „Epistemology of Biotechnology" is appeared in German language as „Epistemologie der Bio –und Gentechnologie" (pp. 285-297) in Klaus Kornwachs (Hg.) Technik – System – Verantwortung, LIT Verlag, 2004 (Technikphilosophie Bd. 10).

Another version of Chapter 9 „Visions of Technology" is published in the Ubiquity: Volume 8, Issue 10 (March 13, 2007 - March 19, 2007).

Acknowledgments

Contents

Paradigmatic Shifts in the Contemporary philosophy of technologies: Culture of technological reflections 13

    I. Introduction to contemporary philosophy of technologies 13
    II. Technics is Embodied and Hermeneutics: New Models 23
    III. Expanding Phenomenology of Technology 32
    Cited and Non-cited References 39

Social and Ethical Aspects of Biotechnological Practice 47

    I. Structuring of Scientific – Technological Innovation 47
    II. Application of Biotechnology in medicine and Agriculture 49
    III. Concluding Remarks 52
    References 53

Technological Development and Social Progress 55

    I. Introduction 55
    II. Conceptions of the technical progress in the perspective of
    historical philosophy of the Enlightenment 56
    III. Theoretical concepts of technological development
    since the Industrial Revolution 57
    IV. Models of technological development 58
    V. Technical and social progress 60
    References 62

Technology Transfer and Modernization: What Can Philosophers of technology Contribute? 65

    I. Technological Development as Modernization 66
    II. Culture and Technology Transfer 69
    III. Technologization and cultural development 71
    IV. New Concepts of Technological Modernization 73
    V. Suitable technology and cultural identity in developing countries 75
    References 76

Ethical Action in Robotics 79

    I. Introduction 79
    II. Perspectives of Ethical Acts 83
    III. Autonomous Robots 86
    IV. Decision Making 89
    V. Conclusion 90
    References 92
    Note 92

Epistemology of Biotechnology 93

I. Laboratory science as the constitution of a Research Practice 93
II. From the Mendelgenetics to Synthetic Biology:
Technologisation of a Research Practice 97
III. Methods and Technical Procedures of the Genetic Engineering:
Technologisation of a Laboratory Practice 101
IV. Synthetic Life - Perspectives towards Genetic Engineering
as a Science of Construction of Organisms 104
References 105

Justified Trust in Technology 107

I. Trust in technology through the scientification of construction? 108
II. Maintenance and technical security:
institutional reason and foundation for trust in technology? 108
III. Limits of the controllability with modern technology 109
IV. Technology at the forefront 110
V. The „revenge" of technology (Technological Revenge) 111
VI. Risk and technical acceptance 112
VII. Innovation and routine: Technical trust due to experience 113
References 115

Postphenomenological Inquiry into Brain Research and Human-embodied Mind 117

I. Naturalization under inclusion of the observer perspective in empirical sciences 117
II. The observer and perspective: A new paradigm for neuro-philosophy 118
III. Postphenomenological analyses of human-embodied mind 120
IV. The central meaning of different forms of memory 121
V. Observer-centered interdisciplinary neuro-philosophy 122
References 125

Visions of Technology 129

Epilogue: Questions Concerning Technology 135

Appendix: Bibliography of Bernhard Irrgang´s philosophical Works 137

I. Monographies 137
II. Series Editor 139
III. Joint Authorship 139
IV. Joint Editorship (Books) 140
V. Papers on the Ethical Discussion 140
VI Papers ont the Philosophy of Technology 145
VII. Systematic Treatise and Essays 150
VIII. Essays on the Historical Problems Investigations 151
IX. Course Material 152

Paradigmatic Shifts in the Contemporary philosophy of technologies: Culture of technological reflections

Arun Kumar Tripathi

I. Introduction to contemporary philosophy of technologies

Today, it is even harder to imagine a life and a day without complex technological systems of energy, transportation, waste management, and production, each of which has at its core a theoretical deductive model as well as practical rules to identify, or to construct and exemplify real situations to which the theoretical models apply. Our real lifeworld is mostly a constructed environment, built around technologies and technological systems that provide the background, context, and medium for human living. Contemporary life is a „technologically mediated life." Since the humanities represent a source of fundamental human skills necessary to give meaning and direction to our technological age, and the study of humanities in technology will show the interconnection between technology, engineering and the arts. Our work on the philosophy of technology - as an exercise in cultural hermeneutics and ethical hermeneutics - is a plea that the task of the philosophy is to work out suggestions concerning basic cultural and ethical conditions of technological and economic development. In principle, a philosophy of technology is concerned with fundamental questions concerning the proper understanding of a technology; how it affects human existence and reciprocally how human existence affects the technology (Irrgang 2008 & Kaplan 2008). The development of a philosophy of technology is, in principle, based on the assumption that substantial philosophical questions can be posed against the technology as proposed=or in the view of the social implications arising out of new organizational, economical and technological developments such as globalization, economics, population growth, ecological crisis, north-south conflict, world-wide communication technology and information distribution. Thus the relevant questions posed by a philosophy of technology are: „Have we access to the techniques or technologies that we need?" „Do we need the technology that we have?" The answers to these questions affect in the long run everybody and everything (Irrgang 2008).

We rely on what we make in order to survive, to thrive and to live together in societies. Sometimes the things we make improve our lives, and sometimes they make our lives worse. Technological devices shape our culture and the environment, alter patterns of human activity, and influence who we are and how we live. A philosophy of technology is a critical, reflective examination of the nature of a technology as well as the effects of developing such technologies upon the public resources of knowledge, as well as on social activities and environments (Kaplan 2009 & 2007). The goal of a philosophy of technology is to understand, evaluate and criticize the ways in which technologies reflect as well as change human life individually, socially and politically. It also examines the transformations effected by technologies on the world of nature, biological life and its ecospheres. The assumption underlying a philosophy of technology is that the devices we make and use and substances we produce and apply transform our experience in ways that are philosophically relevant. That is, technology not only enlarges and extends our capacities to change the natural and social worlds but also does so in ways that are interesting with respect to fundamental areas of philosophical inquiry. Technology then poses unique practical and conceptual problems for epistemology, metaphysics, moral philosophy, and political philosophy. The task for a philosophy of technology is then to analyze the phenomenon of technology, and the ways it significantly mediates and transforms our experience of the lifeworld (Kaplan 2008).

Coming from the school of critical theory in Frankfurt (associated with such scholars as Jürgen Habermas, Adorno, Herbert Marcuse and Horkheimer); Andrew Feenberg proposes a solution to the problems of a philosophy of technology from political perspectives. Currently Feenberg is the most productive philosopher in the area of technology and politics (From a European perspective and one of Technology as Power, cf., Bernhard Irrgang's, Versuche über Politische Technologie). Feenberg does not hesitate to lay bare the skeleton of his argument in clear and helpful charts in his Questioning Technology (Routledge, 1999). Over the course of more than two decades, Andrew Feenberg has established himself as an important representative of a new generation of critical theorists[1]. Consistently insightful and articulate, Feenberg has developed a trenchant critique of technological culture that has taken as its point of departure the humanistic Marxism of his mentor Herbert Marcuse. In Questioning Technology, Feenberg presents what is arguably his most successful attempt to date to construct a major revision of the critique of technology advanced by Marcuse and other „first generation" critical theorists, as well as by their „second genera-

---

1   See Larry A. Hickman. From Critical Theory to Pragmatism: Feenberg's Progress in Democratizing Technology: Andrew Feenberg's Critical Theory of Technology, SUNY, 2006. For a fruitful exchanges between Feenberg (Pragmatism and Critical Theory of Technology) and Hickman (Revisiting Philosophical Tools for Technological Culture) see Techne essays at http://scholar.lib.vt.edu/ejournals/SPT/v7n1/feenberg.html and http://scholar.lib.vt.edu/ejournals/SPT/v7n1/hickman.html

tion" heirs, such as Habermas. Feenberg's new book is: Heidegger and Marcuse: the Catastrophe and Redemption of History (Routledge, 2005).

Feenberg argues against both essentialism and determinism – to put forward a political theory of technology which embraces the social dimensions of technological systems, including their impact on the environment and workers' skills and their role on the distribution of power. Feenberg wants to encompass the technical dimension of our lives and to provide a social account of the essence of technology which enlarges our democratic concerns. On technical democracy, Feenberg reminds us – that a technological society requires a democratic public sphere sensitive to technical affairs. But it is difficult to conceive the extension of democracy to the management of technology through procedures such as voting. Nevertheless, local publics do become involved in protests over technical developments that cause them concern. Hence the widespread recourse to protests and public hearings in domains such as care for the environment – we are witnessing the slow emergence of a technologically concerned public sphere that has been largely overlooked because its concerns are unfamiliar and fragmented.

Andrew Feenberg rejects extreme standpoints on technology. He opposes „substantivism" (Ellul, Heidegger, and Weber) for claiming the only way to deal with the dominance of technology is to oppose it. Ditto for „essentialism" (Borgmann), which says technology has an immutable essence outside history and is beyond our intervention. He also rejects technocratic determinism, which always sees the latest stage of technological development as inevitable and leading us straight to freedom and happiness.

Instead, Feenberg proposes „constructivism." This approach and its accompanying „innovative dialogue" affirm „the social and historical specificity of technological systems, [and] the relativity of technical design and use to the culture and strategies of a variety of technical actors." Technology is thus neither neutral nor autonomous but ambivalent: It is always open to „alternative developments with different social consequences." Feenberg seeks a radical democratic politics, a transformation whereby the „social control of technology will eventually spread and be institutionalized in more durable and effective forms." The process of „deep democratization" will recognize participant interests through „electoral controls on technical institutions."

The most effective way to silence criticism is a justification on the very terms of the likely critique. When an action is rationally justified, how can reason deny its legitimacy? This paper concerns critical strategies that have been employed for addressing the resistance of rationality to rational critique especially with respect to technology. Foucault addressed this problem in his theory of power/knowledge. This paper explores Marx's anticipation of that approach in his critique of the „social rationality" of the market and technology. Marx got around the silencing effect of social rationality with something very much like the concept of under determination in his discussion of the length of the working day. There are hints of a critique of technology in his writings as well. In the 1960s and '70s, neo-Marxists and post-structuralists demanded radical changes in the technological rationality of advanced societies. Soon technical controversies spread, primarily through the influence of the environmental movement. The concept of underdetermination was finally formulated clearly in contemporary science and technology studies, but without explicit political purpose. Nevertheless, this revision of the academic understanding of technology contributes to weakening technocratic rationales for public policy. A new era of technical politics has begun. (Cf. Feenberg, Andrew. Marxism and the Critique of Rationality: From Surplus Value to the Politics of Technology." Forthcoming).

In brief, Feenberg's intent was to respond to the silencing of critique by invoking rationality. He asks: „When an action is rationally justified, how can reason deny its legitimacy?" If it's rational to receive a good in exchange for money, how could there be anything wrong with our capitalist society? Never mind that within that seeming equivalence of exchange, one class is continually enriched, while the other barely holds its ground.

Feenberg finds a way around this silencing of critique in Marx's method (clearly distinguished from the content of his theory), which anticipates Foucault's power/knowledge formulation. The nascent concept of underdetermination in Marx emerged more fully in contemporary science and technology studies, despite its apolitical aspect. According to Feenberg, „this revision of the academic understanding of technology contributes to weakening technocratic rationales for public policy. A new era of technical politics has begun."

Recently Feenberg (by recalling his plea from Questioning Technology in Technical Democratization) put the best way on how the phenomenon of democratizing technology „would be" possible and if we assume, it happens then what will be its consequences, as we all know technology has the Janus-Face. In the book „Five Questions in Philosophy of Technology" edited by Evan Selinger and Jan Berg Olsen (Automatic/VIP Press 2007), Prof. Feenberg has answered the issues on „practical socio-political obligations follow from studying technology from a philosophical perspective:" '[Andrew Feenberg writes that] the main obligation philosophy of technology teaches are responsibility for our own creations and for the consequences of our own actions. We know we should take such responsibility in personal affairs, but what about our relation to nature and to society? Most of the institutions and ideas that we received from the past tell us the natural world is a vast public grab bag and a vast public garbage dump for which we declaim any responsibility. As for society, we are told that our responsibilities begin and end with paying taxes and voting. These are catastrophic errors. Technology is a collective project of society as a whole and can only be brought within the scope of our ethical obligations through a wide variety of political interventions, including protests, boycotts, and active collaboration with experts around new visions of the technical future.'

This is the reason why Feenberg is most concerned with the implications of technology for democracy, a subject that is still largely overlooked. Technologies form the framework of our lives but they are designed with little or no democratic input. This is a serious failure of our institutions, Feenberg says, it must be addressed by reforms in education, the media, the corporations, law, and the technical professions.

Later in the same book, Feenberg writes that idea of democratizing technology has many sources. Perhaps the two most important philosophers to advocate this idea were Marx and Dewey. Marx believed that worker control of the factory could transform modern society and the technology on which it is based. Dewey also hoped for wider citizen participation in technological decision-making. Here is the critique; neither [Marx nor Dewey] had significant examples of democratization to point to. Furthermore, technological determinism was far more popular than their democratic position until quite recently. Indeed, Marx was understood for generations as a determinist. In recent years this has begun to change due to democratic interventions into the management of technology by users and victims and to frequent calls for alternative technologies from scientists and technical experts in fields such as environmental protection and medicine.

In his works after Being and Time, Heidegger argues that the Cartesian, science-centered, enlightenment outlook that has become the dominant intellectual mood in the developed Western world is neither Cartesian nor even modern in origin. Rather, this outlook understands modern science and its applications as a kind of culminating event, an occurrence that fulfils the quest for cosmic knowledge which has inspired Western philosophy from its start. According to Heidegger, this explains why, epistemologically speaking, philosophy is now so widely seen as most see it when it analyses/defends only what is not incompatible with scientifically informed knowledge, and why, ontologically speaking, the world tends to seem most „real" to the extent that it is controllable by modern technoscience. To „be" today seems above all to „be in" this world.

Most contemporary philosophers of technology such as Don Ihde, Luce Irigaray, Hubert Dreyfus, Andrew Feenberg, Robert Scharff, and Bernhard Irrgang would probably agree that, for good or ill, Heidegger's interpretation of technology, its meaning in Western history, and its role in contemporary human affairs is the single most influential position in the field. Much recent philosophical study of technoscience can be seen as a reaction to his interpretation. The reactions are of three kinds. First, Scharff and Feenberg respond by working out the ontological, epistemological, and socio-political consequences of a basically Heideggerian viewpoint. A second viewpoint is to turn away from such „global" considerations and explore in greater detail what it is to „be with" technology-- using in the process Heidegger's own position to disclose technoscientific riches in life that he has ignored or to berate him for being too negative about being-with-technology. A third viewpoint, according to Scharff, is to develop political and social programs that might help combat the often oppressive features of current technoscientific existence; this is what Heidegger at the end of his life called a „free relation" with technology.

From a Euro-American perspective technology is viewed through its connection with the sciences, while in South America the perspective is the reverse, science is viewed through its technologies understood as cultural instruments; this places the technification of sciences in the foreground. Don Ihde as a representative of the North American phenomenology of technology and on the other hand Bernhard Irrgang as a representative of the German phenomenology of technology would like to interconnect both traditions. Ihde understands technological development in terms of a social anthropology of technosystems on the analogy of ecological systems --or as a technologically arranged ecological system. This viewpoint is in opposition to the technological determinism of applied natural science or the determinism of pure technological development (Ihde 1990, P. 5). But these accounts are based on the incorrect notion about technological development that it takes place without any context, whereas in fact the phenomenological underpinning of a technology has an impact on the cultural environment of technological development. The philosophy of phenomenological analysis examines the horizon of human-machine (human-technology) conscious coordinate activity.

Dresden phenomenologist philosopher of technology Bernhard Irrgang (Vol. I: Technological Culture, Instrumental Understanding and Technological Action; Vol. II: Technological Practice, Design Perspectives and Technological Development; Vol III: Technological Progress, Legitimation Problems and Innovative Technology) introduces the thesis of a phenomenological and hermeneutics point of view within the philosophy of technology. Based on the problems in scientific theory and technological sciences, and based on the concept of technological action and implicit knowledge, Irrgang uses a concept of the development of technological know-how (technische Koennen) as a foundation for the meaningfulness of knowledge -- which deals with social, institutional, cultural and ethical elements in society. In the center of the study, a philosophical reconstruction of technology within historical perspectives is developed. Thereby, question about technological and social progress is examined. Based on the concept of technological action and a hermeneutics of technological construction, Irrgang brought these two aspects together with social examples and the analysis of technical institutions. In his works, Irrgang has evaluated the philosophy of technology within the hermeneutics and phenomenology of technology.

The approaches of innovation culture and technology transfer as cultural transfer cannot be conceptualized only historically-institutionally, but must also be done terminologically-methodically. A path of technological development is formed by tradition and innovation. Often it describes a certain shift after a phase of technological progress. However, frequently enough it is connected with visions of progress, at least of the technological means. Speed of innovation differs and depends on cultural factors. Acceptance, cultural assimilation and the interaction of technological paradigms are necessary preconditions for standardisation processes and successful technology transfer. The enforcement of a paradigm requires co-operation and co-ordination. Technology transfer without appropriate cultural transfer is not sufficient: it produces more environmental problems than it possibly avoided. Technology transfer also changes the basic cultural conditions of a society. Heteronomous cultural transfer encounters culturally motivated resistance or neglect. Technology transfer does not automatically lead to modernization, but to forms of development that are culturally adjusted. This process can be mastered by taking the embedding paradigm into account (Irrgang, 2006). At this junction, the processes and paradigms are to be analyzed in the proposed project. Adapted technology is a social and cultural status that is not inherently present in technology. Therefore, technology must be modelled on certain culturally shaped ideals of security, on ideals of the user or environment. However, handling is a cultural evaluation criterion, which is frequently shaped by prejudices (e.g. concerning users) or by once own conceptions of security and environment. These unconscious prejudices and cultural orientations have to be admitted, reflected and discussed. This is the main task of technology reflection culture (Irrgang, 2002a; Irrgang, 2002b, Irrgang 2006).

In order to address practical questions in philosophy of technologies, philosophers such as Hans Lenk, Walther Zimmerli, and Bernhard Irrgang have been developing a hermeneutic understanding of both technology and ethics. The structures of technological practice, professional activity, and everyday life, together with the background of an implicit technological knowledge, are the basis of collective technological action in a cultural context. The meaning of a technology does not necessarily have to be linguistically articulated in order to be present in a culture. The ways technological practices themselves structure actions include different forms of meaningfulness. This leads to a kind of existential pragmatics of technological action and its models of representation. Such an approach provides a recursive and reflexive assessment of technological actions. But the impacts of any interpretation of technological actions must also prove successful in psychological, sociological, technical-historical, and cultural-historical terms. At the same time, reflective modernization depends on the continued existence of such institutions as universities and research centers even as they are altered by globalization. Reflective modernization must also distinguish the self-understandings of scientific and technical professionals from the external descriptions of their roles. The traditional epistemological foundation for a social role description has been the notion of science as knowledge, but technological science is not another science. A metatheory of the technological sciences is needed to determine the relation of these various disciplinary formations and to search for unity within the technological sciences. A related question concerns the relation between disciplinary, interdisciplinary, and transdisciplinary technoscientific knowledge. Epistemological and professional distinctions ultimately interact with practice-orientated and institutional differentiations in an integrated technology-reflective culture.

Don Ihde and Bernhard Irrgang in his work on the Philosophy of Technology argue to the effect that technology, rightly viewed, i.e. phenomenologically understood, is an essential of socio-historically situated human nature. It is basically cultural articulation of man and not an external adjunct. Ihde then proposes a thesis of technology transfer is in effect a sort of culture transfer. Materiality of technological culture does not negate its cultural or human underpinnings. Therefore, whenever some form of technology, agricultural or metallurgical is transferred by way of import of export it carries with it a whole set of human relationships. Transfer of technology is to be understood as a sort of intercultural encounter and gradual accommodation, not confrontation. Differences of culture promote and provides for mutual learning and not necessarily entailing clash and conflicts[2].

„The philosophy of technology is a special region of inquiry. On the one hand, it is continuous with other philosophical topics. For example, practitioners of the philosophy of technology defend their research by appealing to both instrumental and intrinsic justifications—that is, they emphasize how their analyses clarify what it means to be human, and portray alternative visions of how humans and non-humans can relate to each other. On the other hand, the philosophy of technology revolves around unique themes and unorthodox methods. A window into these can be found in the following prefatory remarks" (See Philosophy of Technology: 5 Questions, Edited by Jan-Kyrre Berg Olsen & Evan Selinger Automatic Press / VIP, February 2007).

Don Ihde as elsewhere[3] argued that contemporary philosophy of technology has arisen and grown out of the 'praxis' traditions, particularly those of a concretist orientation, and thus stand in contrast to the earlier, dominant strands of a theoretically biased philosophy of science. And, even if much contemporary phi-

---

2 Along the similar line of ideas, Professor Hans Poser (TU Berlin) in (1991-93) in the papers Die kulturelle Vielfalt und die Förderung wissenschaft-technischer Innovationen and Technology Transfer and Cultural Background argues for technology transfer as a culture transfer. Other publications of Hans Poser where he discussed the perspectives of philosophy of technology and technological culture are: Perspektiven einer Philosophie der Technik in Allgemeine Zeitschrift für Philosophie, 25 (2000), 99 – 118 and Vernunft and technische Kultur in Technikkultur. Von der Wechselwirkung der Technik mit Wissenschaft, Wirtschaft und Politik, Berlin: Stiftung Brandenburger Tor 2002, 162 – 178.

3 See P. Kemp (ed.) World and Worldhood, 91-108, 2004.

losophy of science has been late to arrive at such praxis phenomena as experiment, instrumentation and technologization, in science, it, too, has begun to take a similar direction. This has some implication for the role of the philosopher of technology or of technoscience as current coin would have it.

First, there is some degree to which the philosopher of technology must go native, by this Ihde says it become more than a distant observer, to become an informed participant. Without this participant-observation, the philosopher could never deal with the developmental phases of technologies, which Ihde has argued are as, if not more, important than the response phases which deal with already extant technologies and their effects.

Second, a praxis orientation is necessarily more 'pragmatic' and area or regionally focused than a high altitude and general theory might be. With this focus Ihde sees nothing wrong with focused specialization directed towards the various areas of the technologies of the times.

Third, as indicated above, a classical role for philosophers of technology remains conceptual in the sense of re-conceiving or redescribing phenomena. In this sense one positive feature arising from postmodern sensibility is the appreciation for alternative frameworks and the fusing of horizons in a Gadamerian fashion.

Don Ihde, Peter-Paul Verbeek and Bernhard Irrgang as well plead that philosophy of technology is necessarily concretist and 'materially' oriented insofar as the technologies operate materially at whatever level. Such material operations, as they conclude display patterned, structured, and while multistable, limited sets of possibilities. It is this structure that philosophers in „R & D" may examine and analyse. All of this characterizes a certain style of philosophical approach which is beginning to show itself in the new sub-field of the philosophies of technology.

We require an understanding of science and technology on the basis of culture, wisdom, ecology and ethical values. Today, more than ever before, Irrgang argues there is an urgent need to understand the global imperative of modernization and its atttendent idiom of globalisation. The process of current globalisation is emerging into a cultural, historical, ecological phenomenon. At the same time, this change is adding an ethical dimension to the development of technology, which has an orientation to the understanding of techniques, technology and science. In the last thirty years our world has seen the emergence of cultural understanding of technology and scientific knowledge. These developments are inspired from the American philosophy of technology and continental phenomenology. Their understanding of technological action as the basis of implicit knowledge and motivated by Martin Heidegger's understanding of technical action as an acquaintance with >>Zeug<< (Heideggerian terminology) which are developed into a cultural-institutional understanding of technology which allowed and formed into a new shape and design of technology. This has become the foundation of technology assessment in philosophy, technology and ethics research (Irrgang, 1996; Irrgang, 2001a).

Later, system-theory analysis (employing cybernetics to control technology) has given us a model of social anthropology of technological and cultural development in technological practice. Thus, we can see an adaption and processing of nature as the resource. The development of population, urbanisation and the development of technical institutions can be seen as an esteemed and distinguished central determination of a component of technological development. In the center of our research, we can perceive the reconstruction of industrial revolution as an essential phase of technological development in the two phases: changes of working organization by the use of implementation of implicit technological knowledge in the areas of textile industries and the changes of resource basis by the use of conversion of coal as an energy medium. The central analysis and anatomical artefact is also the integration of technological understanding into everyday life. Thus, changes coming from mass production and the consumer society in the industrial civilization can be witnessed. In the center of study, we have questions of transcultural technology-transfer, eco-social technological modernization and the development of scientific theory of technological sciences and technology. Also in this area, the understanding and meaning of societal issues, for example works, as the guiding principles for technological construction of artefacts can be seen in terms of the conceptual design of technological expression and formation of technological and ethical values (Irrgang 1998; Irrgang, 2002a, Irrgang, 2002b, Irrgang 2007).

In the contemporary philosophy of technologies questions such as how human behaviours and embodiment is affecting the social and cultural factors? How we relate to technologies in the lifeworld? Which kind of relationship do we stand to technologies? And how lifeworld shapes Technology and technology shapes the lifeworld? are playing a key role. Human experiences of our lifeworld are shaped by physical and symbolic tools and mediating tools. A common denominator in the design of many „innovative" learning environments is the insightful and careful application of computer based measurement technology as a mediating tool. Tools are a means of controlling and steering the interconnections between things and a device for coordinating shared human activities. One quote[4] from the 1938 Logic by John Dewey, which clearly says retooling requires retooling the culture: Tool and utensil, every improvement in technique, makes some difference in what is used and enjoyed and in the inquiries that arise with reference to use and enjoyment, with respect to both significance and meaning. Changes in the regulative scheme of relations within a group, family, clan or

---

4   I am grateful to Jim Garrison (Philosophy of Education, Virginia Tech in Blacksburg) for pointing me this quote of Dewey.

nation, react even more intensively into some older system of uses and enjoyments. (John Dewey: Later Works.12.70).

Human ====> Artifact =====> World
Human <==== Artifact <====== World

Above phenomenon is called as the role of artifacts as a mediating tool in human perception. Philosophy of technology deals with such questions (see above and below) as what role does technology (artifacts) play in everyday human experience:
• How do technological artifacts affect the existence of humans and their relations with the world and within our world?
• How do artifacts produce and transform human knowledge and how is human knowledge included in artifacts?

The North American phenomenologist philosopher of technology Don Ihde (see his book „Technology and Lifeworld: from Garden to earth") has developed the following schematic distinctions regarding the intentional relationship between humans and their world:

Embodiment relations:

(Human <=> Technology) <=> World

Hermeneutic relations (hermeneutic orientation to the world):

Human <=> (Technology <=> World)

Alterity relations:

Human <=> Technology (<=> World)

II. Technics is Embodied and Hermeneutics: New Models

In embodiment relations we are not normally aware of the technology. In hermeneutic relations some kind of interpretation is involved, hence the term hermeneutic. Both in embodiment and hermeneutic relations experience is transformed by the mediating technology used. The way technologies are implemented in the relation Human ⇔ Technology ⇔ World shape figure –> background relations.

Persuasive interfaces in a class of interfaces[5] belonging to a trend in contemporary Human Computer Interaction (HCI) where user experiences matter more than for instance user performance. Fallman paper[6] argues that in this shift there

---

5   European human-computer interaction researchers are trying to create tangible interfaces that will take computer interaction possible via augmented physical surfaces, graspable objects, and ambient media such as walls and tabletops. The goal is to make interaction natural and eliminate the need for a handheld device. The Tangible Acoustic Interfaces for Computer Human Interaction (Tai-Chi) Project is currently researching acoustics-based remote-sensing technologies with the goal of transferring information pertaining to an interaction by using the structure of the object as the transmission channel. According to Tai-Chi project users will be able to communicate freely with a computer, interactive system, or the cyber-world using, everyday objects. The project is developing different methods of sensorimotor skills for contant-point localization. Efforts include utilizing the location-signature embedded in the acoustic wave patterns caused by content, and triangulation and acoustic holography. The project's goal is to develop acoustics-based remote-sensing technology that could be adapted to physical objects to create tangible interfaces. Visit http://www.taichi.cf.ac.uk/ for the details. On HCI Design, see Fallman, Daniel. (2003) In Romance with the Materials of Mobile Interaction: A Phenomenological Approach to the Design of Mobile Information Technology, Doctoral Thesis, ISSN 1401-4572, RR.03-04, ISBN 91-7305-578-6, Umea University, Sweden: Larsson & Co:s Tryckeri. [Fallman PhD thesis deals analytically and through design with the issue of Human Computer Interaction (HCI) with mobile devices; mobile interaction. Specifically, it is an investigation into and a capitalization on the multistable kinds of relations that arise between the threefold of human user, artifact, and world, and how dealing with this kind of technology and these relations in many ways must be regarded as different from mainstream HCI. This subject matter is theoretically, methodologically, and empirically approached from two to HCI unconventional outlooks: a phenomenological and a design-oriented attitude to research. The main idea pursued in this work is that while HCI for historical reasons follows a tradition of disembodiment, its opposite—embodiment—needs to come into view as an alternative design ideal when dealing with mobile interaction. The tradition of disembodiment in HCI, how it is applied within mobile interaction, and the conceptual switch in focusing on embodiment and human, technology, world relations are thoroughly analyzed and discussed. A proper understanding of these issues are seen as necessary for the primary purpose of this book: to provide designers of mobile interaction with the conceptual means needed to construct new and better styles of mobile interactions.]

6   Daniel Fallman. (2007) Persuade Into What? Why Human-Computer Interaction Needs a Philosophy of Technology, Persuasive 2007, Second International Conference on Persuasive Technology (April 26-27, Stanford University, Palo Alto, CA): Springer & on information technology interface see Croon Fors, Anna (2006) Being-with Information Technology: Critical explorations beyond use and design. PhD-Thesis, Umeå University, Sweden: Department of Informatics (In this thesis theoretical exploration concerning the significance of information technology in everyday life is conducted. The main question advanced is how the reflexive nature of information technology can be envisioned. By this question attention is directed to transformative, experiential and dynamic qualities of information technology, i.e. unknown mergers of information technology and human experience. It is tentatively suggested that it is within everyday life

is also a shift in accountability, but that this shift tends to remain implicit in HCI. What makes a good user experience? To deal with these issues, it is argued that Human Computer Interaction[7] needs to develop a philosophy of technology. Don Ihde[8] argues for a phenomenology of relations between human users, artifacts, and the world and technologies are seen as inherently non-neutral, whereas Albert Borgmann argues that we need to be cautious and rethink both the relationship as well as the often assumed correspondence between what we consider as useful and what we think of as good in terms of technology.

In his Technics and Praxis Don Ihde explicitly emphasized the necessity of a social embedding of technology and science, as Hans Lenk[9] and Günter Rophol[10] did independently in the seventies including what Ihde calls (social) „praxis" as well as a new interpretation. Ihde did more comprehensively emphasize the technological embodiment of science in a literal sense, not only but notably also in „its instrumentation" seeing „a crucial difference" between modern and ancient science (Ihde: 1979, 1991). Ihde epitomizes „the focal point at which instrumental realism emerges" as being „the simultaneous recognition of what I have called the technological embodiment of science, which occurs through the instruments and within experimental situations; and of the larger role of praxis and perception through such technologies" (1991, 99).

Ihde's Technics and Praxis is an introduction to the phenomenology of instrumentation. Ihde provides many examples of the application of phenomenological analysis to sample tools (e.g. chalk, telephone, telescope, etc.) of technology, which could be important for the students, studying physics under philoso-

---

that such transformative abilities are due to be significant. That is, depending on if and how people experience otherness in their relationships with information technology different strategies and responses of being a part of the whole are achieved).

7   On the issue of HCI Design and Embodiment see also Klemmer, S. R., Hartmann, B., and Takayama, L. How Bodies Matter: Five Themes for Interaction Design. Proceedings of DIS 2006, where authors argue that our physical bodies play a central role in shaping human experience in the world, understanding of the world, and interactions in the world. This paper draws on theories of embodiment — from psychology, sociology, and philosophy synthesizing five themes we believe are particularly salient for interaction design: thinking through doing, performance, visibility, risk, and thick practice. We introduce aspects of human embodied engagement in the world with the goal of inspiring new interaction design approaches and evaluations that better integrate the physical and computational worlds. More details at http://hci.stanford.edu/

8   In Don Ihde's words „phenomenology investigates the conditions of what makes things appear as such" (cf. Don Ihde's Postphenomenology, 1993, p. 133). Ihde suggests a post-phenomenology that is not centered on the subject but on embodiment. With the notion of „embodiment" Ihde problematizes the ongoing interrelation between the active and perceiving body (ort hing) and its environment of action (or use). (cited in Introna, Lucas. Phenomenology in ESTE, p. 1408).

9   Zur Sozialphilosophie der Technik, Frankfurt a.M. 1982: Suhrkamp.

10 mit Hans Lenk (Hrsg.) Technik und Ethik, Stuttgart, 1987, 1989: Reclam.

phy of science. At issues is the relation between the human using tools, and either the tools themselves as they present the world (known as „hermeneutic relations") or the world itself as it is experienced through the tools (known as „embodiment relations"). Ihde diagrams these two situations respectively, as:

Human => (Machine => World) and (Human => Machine) => World.

The aim of phenomenological description is to identify the essential or invariant features of experienced phenomena. Ihde undertakes a phenomenological description of several sets of human-technology relations in order to analyze how technologies often mediate and transform our experiences. A phenomenology of human-technology relations shows that the structural dimensions of technological mediation produce a range of possible experiences.

According to Ihde, when we consider the ways our everyday experience is mediated by technological objects, we find several unique sets of human-technology relations, each positioning us in a slightly different relation to technology. One set of relations Ihde calls „embodiment relations" with devices we use to experience the everyday lifeworld and that, at the same time, alter and modify our perception of the world. (Devices, for examples glasses, hearing aids, writing implements, and the handheld tools.) Another set of Ihde calls „hermeneutic relations" that involve instruments that we read rather than use tools. (Devices, for examples clocks, thermometer, spectrographic devices, and other technologies with visual displays, which must be interpreted to be understood.) A third set is „alterity relations", in which technologies appear as „other" to us, possessing a kind of independence from humans as creators and users. (These devices include things like toys, robots, ATM machine, computer games and visual technologies that we interact with as if they are autonomous beings.) The final fourth set is „background relations," in which technologies form the context of experience in a way that is seldom consciously perceived. (This set of devices includes, for examples the lighting, air conditioning, clothing, shelter, and automatic machines that operate in the background subtly affecting our everyday experience.

Robert Rosenberger[11] while developing philosophy of technologies in neurobiological research argues that advanced technologies (e.g., magnetic resonance imaging, electron microscopy) are beginning mediate between investigators and the brains they investigate. Rosenberger in his research analyzes a device called the „slam freezer" that quick-freezes neurons to be studied under the microscope. Employing insights from Don Ihde's philosophy of technology, work that carefully amalgamates continental philosophy with philosophy of science,

---

11 Bridging Philosophy of Technology and Neurobiological Research: Interpreting Images from the „Slam Freezer", Bulletin of Science, Technology & Society, Vol. 25, No. 6, 469-474 (2005).

Rosenberger draws out the practices of interpretation in slam-freezing research. It clearly shows us that philosophy of technologies do have interdisciplinary approach to understanding scientific methodology sets the stage for further philosophical investigation of research in neuroscience.

The primary research objective of the Extension of the Human Senses group at NASA is to research and develop novel algorithms for modeling and pattern recognition in dynamic non-stationary environments. Our work encompasses all stages of using neuro-electric signals for augmentation including: data acquisition, sensor development, signals processing, modeling, pattern recognition, interface development, and experimentation. The research group specializes in developing alternative methods for human-machine interaction as applied to device control and human performance augmentation. Signal processing environment – EHS has developed a distributed data flow based Signal Processing Environment for Algorithm Development (SPEAD) which is used for all of our studies and is available to our partners. This environment allows for someone to program sophisticated machine learning algorithms by wiring blocks together. These blocks run in parallel on standard PCs and Macs and allow for distributed machines to be used. The Extension of the Human Senses group (EHS) focuses on developing alternative human-machine interfaces by replacing traditional interfaces (keyboards, mice, joysticks, microphones) with bio-electric control and augmentation technologies.

Thomas Hughes draws on an enormous range of literature, art, and architecture to explore what technology has brought to society and culture, and to explain how we might begin to develop an „ecotechnology" that works with, not against, ecological systems. From the „Creator" model of development of the sixteenth century to the „big science" of the 1940s and 1950s to the architecture of Frank Gehry, Hughes nimbly charts the myriad ways that technology has been woven into the social and cultural fabric of different eras and the promises and problems it has offered. Thomas Jefferson, for instance, optimistically hoped that technology could be combined with nature to create an Edenic environment; Lewis Mumford, two centuries later, warned of the increasing mechanization of American life. Such divergent views, Hughes shows, have existed side by side, demonstrating the fundamental idea that „in its variety, technology is full of contradictions, laden with human folly, saved by occasional benign deeds, and rich with unintended consequences." In Human-Built World, Hughes offers the highly engaging history of these contradictions, follies, and consequences, a history that resurrects technology, rightfully, as more than gadgetry; it is in fact no less than an embodiment of human values.

Bernhard Irrgang in his futuristically oriented „Posthumanes Menschsein? (Posthuman bodily existence) Künstliche Intelligenz, Cyberspace, Roboter, Cyborgs und Designer-Menschen Anthropologie des künstlichen Menschen im 21. Jahrhundert" discusses virtually all technological developments which take us beyond humanity, like technological simulations of experience, expert systems, artificial intelligence, robots, implants and prostheses, designer-babies and cyborgs. Posthumanes Menschsein (in this book Irrgang question of notion of embodiment in the posthumanism; is the posthuman embodiment possible?) is a thorough anthropological investigation of posthumanism. Verbeek writes that in a highly sophisticated way, Irrgang analyses these developments in terms of the Cartesian body-mind dualism, the philosophy of corporality, action theory, and the philosophical-anthropological tradition. Irrgang meticulously investigates the boundaries between the human and the technological, and between the human and the posthuman. Irrgang in his works on the philosophy of technologies elaborates the thesis that rather than trying to replace humanity, we should try to cooperate with the posthuman entities we are to create and pleads for an ethics of posthumanism (From a postphenomenological perspective Dutch philosopher of technology Peter-Paul Verbeek is working on ethics of posthumanism and materiality)[12].

Verbeek research interests lies within the philosophy of technology, but later turn to the (meta-) ethics: the possibilities to conceptualize the moral dimensions of technology in ethical theory. Verbeek is interested on the question why it is so hard to conceptualize the moral dimension of artifacts - which he has been trying to relate to the specific direction the Enlightenment took, with its modern subject-object distinction and its 'humanism' in the sense of an absolutization of the human subject as opposed to a world of inanimate objects. Verbeek's dissertation[13] (and later become his book What Things Do) discuss the questions, „What role do artifacts play in our technological culture? Along with Irrgang, Verbeek claims that „Our society is flooded with devices: television sets, cars, microwave ovens, cellular phones. How are all these things affecting us?"

---

12   Cf. Verbeek, P.P. (2006), 'Materializing Morality – design ethics and technological mediation', in: Science, Technology and Human Values, Vol. 31 no. 3 (May 2006), ISSN 0162-2439, pp. 361-380; Verbeek, P.P. (forthcoming 2008), 'Cultivating Humanity: Toward a Non-Humanist Ethics of Technology'. In: New Waves in Philosophy of Technology, eds. Jan-Kyrre Berg Olsen, Evan Selinger, Søren Riis. Palgrave at http://www.utwente.nl/gw/ceptes/research_staff/verbeek/culthum.pdf and Verbeek's recently 2007 VIDI NWO (Netherlands Organization for Scientific Research) the research project on „Technology and the limits of humanity: the ethics and anthropology of posthumanism" at http://www.ethicsandtechnology.eu/research/projects/technology_and_the_limits_of_humanity/

13   Dissertation by Peter-Paul Verbeek, with title „The acts of artifacts: Technology, philosophy, design (Karl Jaspers, Martin Heidegger, Don Ihde, Bruno Latour, Albert Borgmann)" which he completed at the Universiteit Twente (THE NETHERLANDS) in 2000 in Dutch language, and later came out (translated by Prof. Robert Crease, SUNY at Stony Brook into English) under the series of „Philosophy, Science, Technology, and Society" by Penn State University Press in March 2005.

Martin Heidegger advanced two approaches to technology: first, in „Sein und Zeit" (1927; English trans. „Being and Time" 1962), that of technology as an implicit or hidden presence in the human lifeworld; second, after the famous „Kehre" (turn), or „turn," that of technology as a form of truth or revealing. The early Heidegger developed an understanding of (technological) experience in „Being and Time", paragraphs 14-18. In the analysis of human existence as a „being-in-the-world" he discovered the everyday character of engagement with equipment as prior to any theoretical presence of objects. As is implicit in the Greek naming of objects as „pragmata", Heidegger argues that technical praxis is the experiential context from which all science is abstracted. It is more accurate to describe science as theoretical technology than technology as applied science. But this „Being and Time" analysis of human interaction with entities or beings is no more than a moment in Heidegger's larger attempt to understand the „meaning of Being."

Now, turning from the focus on the meaning of Being that predominates in his early work, Heidegger's later thought develops a more explicit philosophy of technology. In „Die Frage nach der Technik" (1954; English trans. „The Question concerning Technology", 1977) he argues that technology is not just a practical engagement with the world but a revealing, reveling, a disclosure or truth about the world. What modern technology in particular reveals is the world as „Bestand", that is, stock or resources subject to human manipulation. The coming upon the world as „Bestand" that is operative throughout modern technology as such Heidegger names „Gestell" or (enframing), the promotion of which is for contemporary human beings not something that they simply choose to use or not but a „Geschick" or (destiny). Like any destiny, however, technology as „Gestell" carries with it both opportunity and danger. The opportunities provided by technology are pervasive in the modern world, but the dangers are more hidden and go deeper than the simple risks so commonly associate with technology, such as the risks of automobile accidents or environmental pollution. The most profound danger is that the disclosure of the world as resource will overwhelm the event of disclosing itself, that the experience of one particular kind of truth will obscure the more primordial truth of Being. The ultimate challenge of modern technology is to be true to the greater human destiny of disclosing in the midst of a technological destiny.

Later, in his book, Verbeek is operating on the Ihdean theses of Expanding Hermeneutics (Don Ihde examines what might he called a material hermeneutics which characterizes much practice within the domains of technoscience. Ihde rejects the vestigial Diltheyan division between the humanistic and natural sciences and argues a type of critical interpretation, broadly hermeneutic characterizes both sets of disciplines. Don Ihde examines what he calls a style of interpretation based in material practices relating to imaging technologies which have given rise to the visual hermeneutics in technoscience studies) and Postphenomenology. In the book, Verbeek analyzes the mediating role of technological artifacts, and developed his own „main theses on postphenomenology" critically discussing the work of Don Ihde on human-technology relationships, of Bruno Latour on humans and nonhumans, and of Albert Borgmann on technological devices and focal things.

Recently Robert Rosenberger argues: „Dutch philosopher Peter-Paul Verbeek synthesizes the history of philosophy of technology in attempt to offer a program for analyzing the ethics and aesthetics engineering design. The objects of our world, he claims, constrain and inform the ways we act. So, important ethical decisions must be made at the moments when we are designing these objects. Verbeek uses the sorts of phenomenological concepts reviewed above to articulate how our relations to technology can be anticipated in ways that will help us to make decisions regarding design and ethics. One of his central case studies regards a design company that concentrates specifically on constructing long-lasting products that customers will want to keep, intending ultimately to produce less total waste." [Rosenberger, Robert. (2007) The Phenomenology of Slowly-Loading Webpages Ubiquity. 8(15)]

Technology surrounds us: millions of homes have digital cable and wireless internet connections; telephones can also serve as cameras, music players, and personal organizers; and everything from stereos to computers grow more sophisticated every year. This, of course, is the technology that most of us encounter and even embrace. But lurking behind these gadgets is an arena in which the topic of technology raises troubling questions. Cosmetic surgery, chemical weapons, and cloning are just some of the more recent examples of the uneasy results of our technological progress, and they remind us that technology is Janus-face; something capable of immeasurable good as well as a test of the limits of human morality and power. Thomas Hughes, the eminent historian of technology and acclaimed author of American Genesis, wrote Human-Built World as a similar reminder, revealing the concept of technology as it was framed historically by thinkers who ran the gamut from horrified to euphoric.

However, I also caution against thinking that the technology alone will bring about the change. The technology only allows us to think of new ways of learning. Just as books require good authors, the new technology will require new kinds of learning design engineers. Professionals will evolve who can take the research from learning theories and blend it with the technologies. It is not a simple or inexpensive task, but we already see some glimpses of what the future may bring. Technology extends our communications ability beyond face-to-face talking. It expands it beyond the printed page and reading to a new dimension. It is building a new and more efficient means of sharing ideas and information among all people.

Exploring such competing perspectives, Human-Built World is a concise intellectual biography of the tool of technology. Drawing on a vast body of work created over the centuries by philosophers and architects, social theorists and web designers, politicians and engineers, Hughes charts the multiple ways that technology has been views; sometimes with elation, sometimes with sceptics; by various thinkers. Technology, as he shows here, has not been a slow and steady march to the ever-increasing complexity and sophistication of objects; it has been the subject of debate for centuries about the human will to create, the inherent danger of progress for its own sake, and the Mephisthelean urge to alter everything from the natural landscape to the daily activities of millions. „In its variety," Hughes writes here, „technology is full of contradictions, laden with human folly, saved by occasional benign deeds, and rich with unintended consequences." Hughes' mission here is to restore to technology these contradictions and unintended consequences, and his Human-Built World is a necessary and original guide that recreates technology as the philosophical, moral, and social dilemma it rightfully is.

Is the promise of technology real this time? Thomas Edison and many others thought that motion pictures would change forever the role of the teacher and learner. Radio was heralded in the late 1920s and 1930s as the saviour of our education system. During World War II Disney Studios developed animated learning systems designed to teach very specific tasks. After World War II overhead projectors and audio filmstrips were to become the meat and potatoes of learning resources. Television seemed to promise that one good teacher could reach the world. As a matter of fact, these innovations not only provided interesting lessons, but people actually learned from them. They have all proven to be effective in the teaching process.

However, even with their record of success they have not significantly changed the patterns of learning and teaching now present in most schools around the world. The effective measures of educational innovations are: 1) Does the innovation increase the master skills of the learner? 2) Can the same level of learning be accomplished in a shorter period of time? and 3) Can a teacher teach more students to the same level of accomplishments? Technologies up until this time have been used as supplemental tools to the classroom. In this respect they are an added expense to regular classroom activities that becomes difficult to justify in cost accounting.

How then can we say that networking and computers will change learning and teaching? Are they just another fad that will fade away like the other learning technologies? If we put provocatively put that new digital technologies have the potential for being very different and challenging because they merge all of the previous resources into one accessible unit, following the classical futuristic perspectives: The new technologies can provide real world simulations; Learning modules can be accessed at anytime and from any place; Virtual teams of learners can work together to solve problems; Learners can work on real world problems and have access to experts; and new technologies can give provide voice-activated dialogues between the learner and the computer.

From a Neo-Heideggerian perspective Montana philosophy Albert Borgmann wants us to pry ourself free and grasp actual reality. With its uniqueness, and great in weight and 'burden' it will command our serious attention, whereas Virtual Reality merely requires our fast-fingered manipulation. Borgmann argues that, the flood of Information today threatens to overflow, suffocate and even obliterate actual reality[14].

In „Focal Things and Practices," Borgmann develops Heidegger's analysis by specifying in greater detail what it would mean to launch a „reform of technology." Borgmann's key distinction is between „things" and „devices." To encounter a thing is to engage it fully and to participate in its „world" – all of the social dimensions of using and experiencing something. A device is merely an instrument for producing a commodity and what the device is for. The device, in principle functions inconspicuously by disburdening us and making a commodity available. Thus, the „promise of modern technology" Borgmann explains is that the use of devices will free us from the misery and work imposed on us by nature and social pressures, and in return will make our lives better by liberating and enriching our experience. However, Borgmann point out that technology has failed to live up to its promise of liberation, because it is silent as to the ends, purposes, and goods that we desire. Like Heidegger Borgmann invites us to see through the pervasiveness and self-reinforcing patterns of technology. But he moves beyond Heidegger – by setting up an original voice Borgmann pleads, to reform technology, we need to revive focal things and focal practices.

III. Expanding Phenomenology of Technology

Historically, the term „postphenomenology" was introduced to signify a revised but thoroughly phenomenological approach to technologies and material culture; it is phenomenology applied to the study of concrete human meaning-making

---

14   See my paper Coping with Innovative Technology; Arun-Kumar Tripathi writes: „The flood of information today threatens to overflow, suffocate and even obliterate actual reality, says the University of Montana philosophy professor Albert Borgmann. The 'lightness' of technological information seems bent on overcoming the 'moral gravity' and 'material density' that real things naturally possess and that demand our mindful engagement. Albert Borgmann is not asking us to abandon technological information. But he is calling us to link it effectively to 'things and practices' that provide for our material and spiritual well-being." Go to http://www.acm.org/ubiquity/views/v7i23_coping.html (SOURCE: ACM Ubiquity: Volume 7, Issue 23, June 20, 2006 - June 26, 2006 The ISSN for Ubiquity is 1530-2180).

practices, particularly to technologies. „Classical" phenomenology—first with Husserl, but including most post-Husserlians, excepting Heidegger—dealt with intentionality (human meaning-making) but was little interested in the technological tools of meaning-making practices. Suggestive hints emerged from Husserl's analysis of writing and from Merleau-Ponty's take on how prosthetic technologies can be used to make old – or are they new? –meaningful practices for disabled persons. In the case of Heidegger, while he was clearly one of the forefathers of 20th century philosophy of technology, his work remained primarily focused upon the general nature of the intentionality transformations of technology-in-general in contrast with the next generation of philosophers of technology who wanted to know how new meanings and functions are made with technologies. Postphenomenology focuses on how human-technological devices affect intentionality through meaning-making practices. But it does so with rigorous scrutiny of particular technologies, rather than technology-in-general as in the earlier 20th century thinkers, including Heidegger. Yet, once philosophy of technology reached its late 20th century state, it had become obvious that praxis oriented philosophies such as phenomenology or pragmatism were better suited than logic- or theory-centered analytic approaches to study the cultural and socio-historical effects of technological transformation.

Postphenomenology continues the phenomenological tradition of a 'world' (inter-relationistic) ontology of objects related to one another and culturally to human subjects   In the case of technologies, for example, humans „invent" technologies; while reciprocally, technologies also „re-invent" humans. Co-constitution is recognized in a relational ontology. But, such relational ontologies are not unique to phenomenology—they are part of the family of pragmatic [e.g., organism/environment] and actor network [e.g., humans and their nonhuman 'props'] ontologies as well.

Embodiment, being a body, is also a constant within postphenomenology. But since bodies are actively perceptual and culturally-historically constituted, postphenomenology must take account of the variations and possibilities of diverse embodiments. Thus, issues of different cultures, gender, politics and ethics are included in postphenomenological analyses.

Variational analyses provide the methodological style of this approach. With technologies, there are multiple ways in which any single technology may be related to users and multiple ways in which each technology is culturally embedded. Variations must also be considered with respect to the complex dimensions which are included in all such phenomena. Variational analysis - more precisely, the study of group-theoretic invariants - provides a rigorous method not found in early pragmatism. Thus postphenomenology can be seen as an adaptation to late 20th-early 21st century philosophic needs and issues, particularly in the context of technoscience and material culture.

According to Ihde classical phenomenology, first under Edmund Husserl, was formulated within a specific historical context in which 'modern' philosophies dominated. The philosophy of this period was 'modern' with its distinctions between „subject/object," „body/mind," „external/internal" worlds, and for Husserl was largely exemplified by Descartes and Kant. For science, Ihde says, the early 20th century philosophers of science tended to characterize science as a largely abstract, mathematized, practice which was primarily theory-driven. And, regarding technology, neither philosophy nor science could be said to be sensitive to the roles of material technologies. Further Ihde contends that while Husserl's phenomenology as a new „rigorous science" attempted to radically challenge these notions. For example, Husserl's Cartesian Meditations challenged and inverted Descartes, and his Crisis challenged the early modern notion of science. Yet, in spite of this the shadows of the modern remained attached to classical phenomenology, which ironically became known as a 'subjectivist' philosophy.

According to Ihde and Irrgang it can be seen, that while there were marked differences between these early philosophers of technology—for example, most of the Europeans were interested in technology-in-general, were mostly critical or took a dystopian attitude, whereas the Americans tended more towards optimism and in some degree were more empirically oriented. One could note that it was the praxis philosophies: Marxism, pragmatism[15], and phenomenology—that developed the interest in the material culture.

---

15   On the issue of Pragmatism I would like to divert the reader's attention to a latest book by Larry Hickman on „Pragmatism as Post-postmodernism." The new book by Hickman has a very intriguing title „Pragmatism as Post-Postmodernism: Lessons from John Dewey" is published by Fordham University Press in December 2007. In his new book, Hickman presents John Dewey as very much at home in the busy mix of contemporary philosophy—as a thinker whose work now, more than fifty years after his death, still furnishes fresh insights into cutting-edge philosophical debates. Hickman argues that pluralistic mix of contemporary philosophical discourse, with its competing research programs in French-inspired postmodernism, phenomenology, Critical Theory, Heidegger studies, analytic philosophy, and neopragmatism, invites renewed examination of Dewey's central ideas. Hickman offers a Dewey who both anticipated some of the central insights of French-inspired postmodernism and, if he were alive today, would certainly be one of its most committed critics, a Dewey who foresaw some of the most trenchant problems associated with fostering global citizenship, and a Dewey whose core ideas are often at odds with those of some of his most ardent neopragmatist interpreters. Here is my primary thoughts on Larry Hickman's wonderful forthcoming book „Pragmatism as Post-postmodernism": I think Hickman's new book is expanding thoughtful ingenious ideas which he has presented in the Chapter „Pragmatism as Post-Postmodernism" and contributed in the Elias Khalil's anthology on „Dewey, Pragmatism and Economic Methodology (Routledge Inem Advances in Economic Methodology, 3, 2004); In that contribution Larry Hickman writes: „Put another way, if we take Ermarth's suggestion, that or postmodernism „there is no common denominator - in 'na-

By mid-century other social science movements, also practice centered, accelerated the interest in technologies and materiality. New sociologies of science challenged traditional theory-biased philosophies of science, and turned to laboratory practices and the role of instrumentation. „Social construction," the „Strong Program," „Laboratory Life," began to formulate a new image of science, which was both more socially, multidimensional and materially produced. Later still, feminist philosophers also turned to technoscience with emphases upon gender and embodiment. Don Ihde in his works on the philosophy of technologies concentrates upon the way phenomenology amongst a second generation of philosophers of technology began to incorporate materiality, instruments, praxis into a more postphenomenological style of analysis and its „empirical turn" (Hans Achterhuis) as recognized in recent publications and conferences. All science or technoscience is produced by humans and either directly or indirectly implies bodily action, perception and axis. Don Ihde together with Bernhard Irrgang is pleading for multicultural origins of technoscience.

Technoscience has gained enormous presence in the contemporary world, culturally, physically and epistemologically. Ihde argues all science in its production of knowledge is technologically embodied. Human embodiment implies bodily action, perception and praxis. Scientific knowledge production grounded in cultural and historical realities will be the basis of multicultural origins of technoscience. According to Ihde, astronomy and associated cosmology are the latest revolution in imaging technologies (in comparison to Irrgang, who claims the revolution of technoscience had started during the period of Harappa, industhal civilization, and Maya cultures). Telescopes mediate human perception in a new way: the embodied observer now takes up a technology which at first is literally located between one's active body and the object observed. The technological limits remained largely isomorphic with human bodily limits, with visual limits, claims Ihde.

---

ture' or 'God' or 'the future' – that guarantees either the One-ness of the world or the possibility of neutral or objective thought," then Dewey's naturalistic metaphysics has already gone beyond that skeptical point – beyond the postmodernist idea that there is no common denominator – to the more mature position that Hickman says, I am calling post-postmodernism. A key feature of this post-postmodernism is his argument that there are, after all, common features of human experience, and that those common features are accessible by means of (what I am calling) a post-postmodernist metaphysics as „the generic traits of existence." Following Larry Hickman's argument „Dewey developed his own moderate form of constructivism, then, and this is one of the senses in which he was a postmodernist." [Yes] From a pragmatist perspective, however, what seems to be missing in postmodernism, and what Dewey provides as a corrective, is a theory of experimental inquiry that takes its point of departure from real, felt existential affairs. And this analysis of the „generic traits of existence" is one of the areas in which Dewey's work shines brightly as what I have termed post-postmodernism.

The Third China Lecture on „Visualizing the Invisible: Imaging Technologies" by Ihde develops a case study in a postphenomenological analysis of imaging technologies. In this Ihde contends that science, in a rigorous and robust sense, has always been technologically or instrumentally embodied, but that as its instruments change, so does its world and our understanding of scientific knowledge. Beginning with a historical thesis showing how pre-modern science occurred in many cultures in pre-history which incorporated perceptual observations with simple instruments; then with new optical technologies in early modernity produced 'modern' science; then, only since the 20th century moved into the high-technology imaging instruments, now produces a postmodern science. This newest science simultaneously takes knowledge production into the realm of that which lies beyond human embodiment—but in the process shows how one must necessarily take embodiment into account. This lecture emphasizes the focus of most current science practice upon visualization and illustrates how the invisible is visualized.

The Fourth China Lecture on „Do Things 'speak'? A Material Hermeneutics[16]" by Ihde extends the example of contemporary imaging technologies to technologies which „give things voices," in a series of variations from analogues to many dimensions of the sensorium. The contention here is that the humanities and the social sciences could benefit from the new imaging revolution by incorporating the recent practices of the natural sciences by applying new instrumentation to humanities and social science questions. By llustrating audio-visually, Ihde develops a limited number of case studies to show how such a material hermeneutics can and does work. One of these examples includes „Otzi, the Iceman," or the narrative of constituted knowledge current about the 1991

---

16   Don Ihde examines what might he called a material hermeneutics which characterizes much practice within the domains of technoscience. Ihde rejects the vestigial Diltheyan division between the humanistic and natural sciences and argues types of critical interpretation, broadly hermeneutic characterize both sets of disciplines. Don Ihde examines what he calls a style of interpretation based in material practices relating to imaging technologies which have given rise to the visual hermeneutics in technoscience studies. Vesprey, 1993, it was at that meeting that Don Ihde first proposed the notion of „expanding hermeneutics." A material hermeneutics is a hermeneutics which „gives things voices where there had been silence, and brings to sight that which was invisible." Such a hermeneutics in natural science can best be illustrated by its imaging practices. The objects of this visual hermeneutics were not texts nor linguistic phenomena, but things which came into vision through instrumental magnifications, allowing perception to go where it had not gone before. One could also say that a visual hermeneutics is a perceptual hermeneutics with a perception which while including texts, goes beyond texts. This local history gives but a small glimpse of the directions Ihde tries to outline in Expanding Hermeneutics. In Expanding Hermeneutics Ihde outlines both a weak program of hermeneutics in natural science, that is, a program of actual and extant practices which can best be understood as hermeneutic practices, and whereas a strong program which is more prescriptive, suggesting ways to radicalize a material hermeneutics.

freeze-dried mummy found in the Italian Alps. Another example is the experimental development of instrument produced data, not into the usual visualization practiced by current science, but into acoustic presentations which have implications for both the social and natural sciences.

Within the Euro-American community of philosophers relating hermeneutics to science there is a considerable disagreement about where hermeneutics may be located. The older traditions hold that hermeneutics apply to and are limited to the social, cultural, and historical dimensions of science. But newer approaches claim that hermeneutics applies to the very praxis of science and to the constitution of scientific objects. Don Ihde sides with the latter perspective and argues that a tendency to retain vestigial positivist interpretations of science keeps the older tradition from seeing hermeneutics as deeply embedded in science praxis. After arguing this point historically, Don Ihde turns to a hermeneutic recuperation of science, first by drawing from the hermeneutic approach of Joseph Rouse, and then by the "hermeneutic" constructionism of Bruno Latour. Ihde finally turns to what he terms "technoconstruction"in science, particularly in imaging processes, to show concrete cases of the hermeneutic preparation of scientific objects. In the end Ihde concludes that contemporary science has exceeded its earlier modernist framework and now operates in a constructionist-hermeneutic framework. [See Don Ihde, Expanding Hermeneutics: Visualism in Science]

In „Why a hermeneutical philosophy of the natural sciences?" Heelan gives impetus upon the necessities to address the philosophic crisis of realism vs. relativism in the natural sciences. This crisis is seen as a part of the cultural crisis that Husserl and Heidegger identified and attributed to the hegemonic role of theoretical and calculative thought in Western societies. The role of theory is addressed using the hermeneutical circle to probe the origin of theoretic meaning in scientific cultural praxes. This is studied in Galileo's discovery of the phases of Venus; the practice of measurement; the different theories and practices of space perception; the historicality and temporality of scientific research communities which ground paradigm change; and the process of discovery.

On the paradigmatic reflections along with Feenberg, Ihde and Irrgang I would also like to introduce the works of Patrick Heelan who is the William A. Gaston Professsor of Philosophy and a member of the Jesuit Order at Georgertown University. Over the years, Patrick Heelan has been specializing in the study of Heisenberg from the perspective of Husserl, Merleau-Ponty, and Heidegger. The main focus of Heelan's research interests include: (i) on perception, and especially, on the geometry of visual perception in everyday life and pictorial art and architecture; this interest continues an interest evoked when studying relativistic cosmology with Erwin Schrödinger, and (ii) on the role of consciousness (subjectivity, intentionality) in research in the natural sciences, with a special interest in studying the hermeneutics and phenomenology of quantum measurement; this last presently unresolved problem was raised by Nobel Laureate Eugene Wigner in the 1960's with whom for two years as a post-doc at Princeton. Patrick Heelan[17] is the founder of a branch of the philosophy of natural science called the 'hermeneutic philosophy of science'. This work springs partly from the phenomenological tradition of Edmund Husserl, Maurice Merleau-Ponty, and Martin Heidegger, and partly from the Bernard Lonergan's critique and development of the work of Aristotle, Aquinas and Kant. By 'hermeneutic', Heelan means, 'interpretation' or the 'meaning-making" in the formation of concepts, judgments, and actions; in particular by perceiving, by theorizing, and by using laboratory instruments in scientific measurements. Don Ihde in his book on „Experimental Phenomenology" (SUNY, 1986) on the perspective of „Interdisciplinary Phenomenology" in „hermeneutics of sciences and instruments mediated perception" writes: One phenomenologically oriented philosopher of science, Patrick Heelan, has argued that the use of instruments modifies the perception substantially. He holds that the „worlds" constituted through direct or mundane perception experience and the „world" developed through scientific instruments are different, the ordinary „world" being constituted by ordinary perceptions and the scientific „world" being constituted by instrument-mediated perceptions (for details see Heelan's paper „Horizon, Objectivity and Reality in the Physical Sciences." International Philosophical Quarterly 7 (1967): 375-412). Don Ihde later in his works on philosophy of technologies has taken note of Heelan's suggestion in a somewhat broader context and perspective and attempted to consider what occurs when experience is directed through, with, and among technological artefacts (machines), of which scientific instruments are a sub-class. Ihde calls this as a phenomenological interpretation of machine- or instrumented mediated experience (for details see Don Ihde's book Technology and Lifeworld, IUP, 1998).

---

17 Dagfinn Follesdal working along the similar lines argues that hermeneutics is often defined as the study of what is meaningful. The meaningful is normally taken to include not only what is written and spoken, but also actions and various human activities, possibly also some animal activities. Follesdal comments that natural science is often contrasted with hermeneutics, but clearly, some aspects of science are studied in hermeneutics, namely he says science as a human activity, and also scientific theories, which have to be understood and interpreted, like all other texts. Follesdal maintains that difference between natural science and hermeneutics is more complicated that it might seem. The phenomena studied in the natural science are not so totally detached from the mental as is usually thought, and that the borderline between hermeneutiucs and natural science is therefore not as clear-cut as it is often held to be. (P. 293, Follesdal, D. Hermeneutics and Natural Science in Fehér, O. Kiss, L. R. (eds.): Hermeneutics and Science. Proceedings of the First Conference of the International Society for Hermeneutics and Science. Boston Studies in the Philosophy of Science, vol. 206, Kluwer Academic Publishers, Dordrecht, 384 pp., 1999).

Heelan has responded to Arun Tripathi's focus (based on personal communication in 2008) on the teaching and learning functions that involve the use of new technologies in the educational process. This is certainly a great concern of society. There are three aspects to technological education: 1) teaching the manipulative skills involved; 2) orienting their use towards beneficial socio-cultural outcomes, and 3) anticipating disruptions to the old rules that hold society together.

However, new technologies lead to a new kind of human being - one embodied in a new technologically enhanced body. Homo is indeed homo faber, and he becomes more so every day. This is the new technologically enhanced human being – who is not an objective artifact (a technology) but a subjective artifact of the new technologically enhanced (perceptually, cognitively, and desire- and institutionally-oriented) human subject. The social/cultural changes that this brings about are usually neither determinate nor generally foreseeable; and, because of this, the changes will demand special oversight. This new technologically-enhanced human being opens up the social imagination of users to new worlds in which there is a redistribution of powers, such as powers to intrude into and manipulate the lives of others with or without their knowledge; powers to snoop, deceive, acquire resources secretly, defeat traditional rights and privileges as well as power to depose existing institutional authorities. Think of the Automobile, Cyberspace and Internet! Their effects are not limited just to the material and informational resources (the technologies as tools) available to a sector of society; they affect the imagination and anticipatory desires of the „new humans." The changes in the making of this new human being are unforeseeable and will eventually demand changes in ethics, laws, social structures, accountability, and institutions (Heelan, 2008).

Consequently, the functions of education vested in those who command and teach the use the new technologies may (should) come under special social scrutiny. Hughes – and not just Goethe of Mumford - makes us aware of the problem of confronting the new and divergent anticipatory social imaginations and desires that can emerge from the re-structuring of the moral and cultural structure of a technologically-embodied society; these give rise to conflicts with moral, philosophical, and religious dimensions that threaten the fracturing of the old institutional order (Heelan, 2008).

Cited and Non-cited References

Achterhuis, H. (Ed.). 2001. American Philosophy of Technology: The Empirical Turn. Bloomington: Indiana University Press.
Ackermann, R.: Data, Instruments and Theory. Princeton 1985: Princeton UP.

Berg Olsen, Jan-Kyrre and Selinger, E. Philosophy of Technology: 5 Questions, edited by, February 2007, Paperback, ISBN 8799101386. Read snippets at www.philosophytechnology.com

Bunge, M.: Scientific Research. 2 Vols. Heidelberg - Berlin - New York 1967: Springer.

Borgmann, A. 1999. Holding onto Reality. Chicago: University of Chicago Press.

Cassirer, E. „The Concept of Group and the Theory of Perception" in Philosophy and Phenomenological Research, V(1944)/1: 1-35, 1944

Corona, N., Irrgang, B.: Technik als Geschick? Geschichtsphilosophie der Technik; Dettelbach 1999 [Technology as Destiny? History of Philosophy of Technology].

Crease, Robert P. (ed.): Hermeneutics and the Natural Sciences, Kluwer Academic Publishers, Dordrecht, 1997 (Reprint of the journal Man and World, Volume 30, Issue 3, July 1997).

Dewey, J. Democracy and Education, in, the Middle Works, 1899-1924, vol. 9, ed. Jo Ann Boydston (Carbondale: Southern Illinois University Press, 1916, 1980).

Dewey, J. 1938. Logic: The Theory of Inquiry. New York, Holt.

Dewey, J. 1957. Reconstruction in Philosophy. Boston: Beacon Press. Enlargement of 1920 edition.

Durbin, Paul T. 1992. Social Responsibility in Science, Technology, and Medicine. Bethlehem, PA: Lehigh University Press.

Durbin, Paul T. 1990. „Introduction: Conflict over Philosophy of Technology as an Academic Field." In P. Durbin, ed., Broad and Narrow Interpretations of Philosophy of Technology. Dordrecht: Kluwer. pp. ix-xvii.

Durbin, Paul T. 1997. „In Defense of a Social-Work Philosophy of Technology." In C. Mitcham, ed., Research in Philosophy & Technology, vol. 16. Greenwich, CT: JAI Press. pp. 3-14.

Durbin, Paul T. PHILOSOPHY OF TECHNOLOGY: IN SEARCH OF DISCOURSE SYNTHESIS Techné: Research in Philosophy and Technology Number 2 Volume 10, winter 2006. http://scholar.lib.vt.edu/ejournals/SPT/v10n2/pdf/v10n2.pdf

Dusek, V. Philosophy of Technology: An Introduction, Blackwell Publishing, 2006. This book complements The Philosophy of Technology: The Technological Condition: An Anthology, edited by Robert C. Scharff and Val Dusek (Blackwell, 2003).

Feenberg, Andrew. Critical Theory of Technology (Oxford University Press, 1991).

Feenberg, Andrew. Alternative Modernity: The Technical Turn in Philosophy and Social Theory (University of California Press, 1995).

Follesdal, D. Introduction and the chapter? Hermeneutics and Natural Science? in Márta Fehér, Olga Kiss and László Ropolyi, eds., Hermeneutic and Science. Proceedings of the First Conference of the International Society for Hermeneutics and Science, Veszprém, Hungary, September 6-7, 1993. (Boston Studies in the Philosophy of Science, Vol. 206) Dordrecht: Kluwer, 2000, pp. 293-298.

Fehér, O. Kiss, L. R. (eds.): Hermeneutics and Science. Proceedings of the First Conference of the International Society for Hermeneutics and Science. Boston Studies in the Philosophy of Science, vol. 206, Kluwer Academic Publishers, Dordrecht, 384 pp., 1999.

Giere, R. N.: Constructive Realism. In: Churchland, D. M. - Hooker, C. A. (Hg.): Images of Science. Chicago 1985: Chicago U P, 75-98.

Giere, R. N.: Explaining Science: The Cognitive Approach. Chicago - London 1988: Chicago Uni. Press.

Giere, R. N.: The Cognitive Structure of Scientific Theories. Philosophy of Science 61 (1994), 276-296.

Giere, R. N.: Science without Laws. Chicago 1999: Chicago U P.

Hacking, I.: Representing and Intervening. Cambridge - New York 1983: Cambridge U P.

Harré, R.: Varieties of Realism: A Rationale for the Natural Sciences. Oxford 1986: Blackwell.

Heelan, P. & Schulkin, J. „Hermeneutic Philosophy and Classical Pragmatism" in Philosophy and Social Action 30 (2004): 23-36.

Heelan, P. „Afterword" Hermeneutic Philosophy of Science, Van Gogh's Eyes, and God: Essays in Honor of Patrick A. Heelan, S.J.. Ed. Babette E. Babich. Boston and Dordrecht: Kluwer, 2002.

Heelan, P. „Why a Hermeneutical Philosophy of Natural Science?" Man and World 30 (1997): 271-298.

Heelan, P. „Quantum Mechanics and the Social Sciences: After Hermeneutics." Science and Education 4 (1995): 127-136.

Heelan, P. „Perception as a Hermeneutical Act." Review of Metaphysics 37 (1983): 61-76.

Heelan, P. „Hermeneutics of Experimental Science in the Context of the Life-World." Philosophia Mathematica 9 (1972): 101-144. [Republished in Don Ihde and Richard Zaner. Interdisciplinary Phenomenology, Nijhoff, 1977]: In this paper Heelan tells us how the lifeworld both of scientific community and of our general culture, is enriched by science and technology and attempt to give a genetic analysis of this enrichment.

Hickman, Larry A. 1990. John Dewey's Pragmatic Technology. Bloomington: Indiana University Press.

Hickman, Larry A. 2001. Philosophical Tools for Technological Culture: Putting Pragmatism to Work. Bloomington: Indiana University Press.

Hickman, Larry A. (Ed.). 1998. Reading Dewey: Interpretations for a Postmodern Generation. Bloomington: Indiana University Press.

Higgs, Eric; Andrew Light; and David Strong. (Eds.). 2000. Technology and the Good Life? Chicago: University of Chicago Press.

Hughes, Thomas P. 2004. Human-Built World: How to Think about Technology and Culture. University of Chicago Press.

Ihde, D.: Technics and Praxis. Dordrecht 1979: Reidel

Ihde, D.: Technology and the Lifeworld. Bloomington-Indianapolis 1990: Indiana UP.

Ihde, D.: Instrumental Realism: The Interface between Philosophy of Science and Philosophy of Technology. Bloomington-Indianapolis 1991: Indiana U.P.

Ihde, Don. Techno-Science and the 'other' continental philosophy in Continental Philosophy; 33: 59-74, 2000, Kluwer Academic Publishers.

Ihde, Don. Thingly heremeneutics/Technoconstructions in Man and World; 30: 369-381, 1997, Kluwer Academic Publishers.

Ihde, D.: Bodies in Technology. Minneapolis-London 2001: U. of Minnesota P.

Ihde, D. and Selinger. E. (eds.) Chasing Technoscience.

Ihde, D.: Imaging Technologies: a Technoscience Revolution/Invited Paper XX World Congress of Philosophy 2003 (Istanbul).

Ihde, D.. (1986) Consequences of Phenomenology. SUNY Press.

Ihde, D. Has the Philosophy of Technology Arrived? A State-of-the-Art Review, Philosophy of Science, 71 (January 2004) pp. 117–131.

Ihde, D. Postphenomenology and the Lifeworld (pp. 39 - 51) in Phenomenology and Ecology, The Twenty-Third Annual Symposium of the Simon Silverman Phenomenology Center, Duequesne University, Pittsburgh, edited by Melissa Geib, 2006.

Irrgang, B. (2003) Von der Mendelgenetik zur synthetischen Biologie. Epistemologie der Laboratoriumspraxis Biotechnologie Technikhermeneutik Bd. 3; Dresden.

Irrgang, Bernhard 2005: Technologietransfer transkulturell. Komparative Hermeneutik von Technik in Europa, Indien und China, Dresdner Studien zur Philosophie der Technologie [Transcultural Technology Transfer. comparative hermeneutics of technology in Europe, India and China]. Peter Lang Verlag, Frankfurt.

Irrgang, Bernhard 1996: Von der Technologiefolgenabschätzung zur Technologiegestaltung. Plädoyer für eine Technikhermeneutik [From the Technology assessment to Technology Shaping. Plea for a hermeneutic of technology]; in: Jahrbuch für Christliche Sozialwissenschaften 37 (1996), 51-66.

Irrgang, Bernhard 1998: Praktische Ethik aus hermeneutischer Perspektive [Practical Ethics from hermeneutics perspective]; Paderborn.

Irrgang, Bernhard 2001a: Technischer Kultur. Instrumentelles Verstehen und technisches Handeln [Technological Culture. Instrumental Understanding and Technological Action, Philosophy of Technology Vol. I]; Paderborn.

Irrgang, Bernhard 2002a: Technischer Praxis. Gestaltungsperspektiven technischer Entwicklung [Technological Practice. Design Perspectives of Technological Development, Philosophy of Technology Vol II]; (Philosophie der Technik Bd. 2); 238 S.; Paderborn 2002.

Irrgang, Bernhard 2002b: Technischer Fortschritt. Legitimitätsprobleme innovativer Technik [Technolgical Progress. Legitimacy Problems of Innovative Technology, Philosophy of Technology Vol. III]; (Philosophie der Technik Bd. 3); 218 S.; Paderborn 2002.

Irrgang, Bernhard, Ricardo Maliandi 2003: Technikphilosophie in Lateinamerika. Themen, Probleme und Entwicklungsperspektiven am Beginn des 21. Jahrhunderts [Philosophy of Technology in Latin America. Themes, Problems and Development perspectives in the beginning of 21st century]; Dresden.

Irrgang, Bernhard. Philosophie der Technik (Philosophy of Technics and Technology), WBG Verlag, 2008.

Kaplan, David. (2009) „Paul Ricoeur and the Philosophy of Technology" in Farhang Erfani, ed. Paul Ricoeur (1913-2005): Remembering a Life, Continuing the Work, Lexington Books.

Kaplan, David. (2008) „How to Read Technology Critically", in J. Olsen, E. Selinger, S. Riis, ed. New Waves in Philosophy of Technology, Blackwell Publishers, 2008.

Kaplan, David (2007) „0Paul Ricoeur and the Philosophy of Technology", Journal of French Philosophy 16, no. 2.

Kuhn, T. The Structure of Scientific Revolutions (1962, with postscript as of 1969). Chicago 1970: Chicago U. P.

Lenk, H.: Philosophie im technologischen Zeitalter [Philosophy in the Age of Technologization]. Stuttgart 1971, 1972: Kohlhammer.

Lenk, H. Pragmatische Philosophie [Pragmatic Philosophy]. Hamburg 1975: Hoffmann & Campe.

Lenk, H. Handlung als Interpretationskonstrukt [Action as Interpretation Construct]. In: Lenk, H. (Ed.): Handlungstheorien interdisziplinär, II, 1. Munich 1978: Fink,279 –350.

Lenk, H. Zur Sozialphilosophie der Technik [Social Philosophy of Technology]. Frankfurt a. M. 1982: Suhrkamp.

Lenk, H. Zu einem methodologischen Interpretationskonstruktionismus. Zeitschrift für allgemeine Wissenschaftstheorie [To a methodological interpretationconstructivism] (Journal for General Philosophy of Science) 22 (1991), 283-302.

Lenk, H. Interpretation und Realität [Interpreation and Reality]. Frankfurt a. M. 1993a: Suhrkamp.

Lenk, H. Macht und Machbarkeit der Technik [Power and Realization of Technology]. Stuttgart 1994: Reclam.

Lenk, H. Einführung in die Erkenntnistheorie. Interpretation – Interaktion – Intervention. Munich 1998: Fink (UTB).

Lenk, H. Kreative Aufstiege. Zur Philosophie und Psychologie der Kreativität. Frankfurt a. M. 2000 (a): Suhrkamp.

Lenk, H. Zur technologie- und handlungsorientierten Wissenschaftstheorie. In: Abel, G. - Engfer, H.-J. - Hubig, C. (Eds.): Neuzeitliches Denken. (Festschrift H. Poser) Berlin - New York 2002: de Gruyter, 61-82.

Lenk, H. Grasping Reality. An Interpretation-realistic Epistemology. Singapore 2003 (in press): World Scientific.

Lenk, H. & Maring, M. (Eds.): Advances and Problems in the Philosophy of Technology. Münster 2001: LIT.

Lenk, H. & Moser, S. (Eds.): Techne – Technik – Technologie. Pullach/Munich 1973: Dokumentation Saur

Lenk, H. & Ropohl, G. (Eds.): Technik und Ethik. Stuttgart 1987, 1989: Reclam.

Mitcham, C. Thinking through Technology: The Path between Engineering and Philosophy. Chicago: University of Chicago Press, 1994. Pp. xi, 397.

Mitcham, C. Thinking Ethics in Technology: Hennebach Lectures and Papers, 1995-1996. Golden, CO: Colorado School of Mines Press, 1997. Pp. viii, 174.

Mead, G. H. 1938. The Philosophy of the Act. Chicago: University of Chicago Press. (A posthumous collection of mostly unpublished Mead papers).

Rorty, R. 1998. Achieving Our Country: Leftist Thought in Twentieth- Century America. Cambridge, MA: Harvard University Press.

Scharff, Robert C. Philosophy of Technology: The Technological Condition, An Anthology, co-edited with Val Dusek. Oxford: Blackwell, 2003.

Scharff, Robert C. „Ihde's Albatross: Sticking to a 'Phenomenology' of Technoscientific Experience," in Expanding Phenomenology: A Critical Companion to Ihde, ed. Evan S. Selinger. Albany, NY: State University of New York Press, 2006. Pp. 131-44.

Scharff, Robert C. „On Philosophy's 'Ending' in Technoscience: Heidegger vs. Comte," in The Philosophy of Technology: The Technological Condition, An Anthology, co-ed. Val Dusek. Oxford: Blackwell, 2003. Pp. 265-76.

Scharff, Robert C. „Technology and Applied Science," A Companion to the Philosophy of Technology, ed. Jan-Kyrre Berg Olsen, Stig Andur Pedersen, and Vincent F. Hendricks. Oxford: Blackwell, 2008.

Scharff, Robert C. Review of Andrew Feenberg, Heidegger and Marcuse: The Catastrophe and Redemption of History. New York: Routledge, 2005. Continental Philosophy Review 40/1 (2007), 91-97.

Scharff, Robert C. „Feenberg on Marcuse: 'Redeeming' Technological Culture," Techné: Research in Philosophy and Technology 9/3 (2006), 62-80.

http://scholar.lib.vt.edu/ejournals/SPT/v9n3/scharff.html

Scharff, Robert C. „Philosophy of Technology," Edinburgh Encyclopaedia of Continental Philosophy, ed. John Protevi. Edinburgh: Edinburgh University Press, 2005. [USA edition: A Dictionary of Continental Philosophy. New Haven CT: Yale University Press, 2006.] Pp. 570-74.

Scharff, Robert C. „Heidegger on Learning to Question Technology," invited paper presented to the Philosophy Department, Stony Brook University, March 14, 2008, Unpublished.

Scharff, Robert C. „How History Matters to Philosophy (cont.): Socrates' Examined Life as a Case Study" presented as a Senior Research Fellow Award paper at the University of New Hampshire Humanities Center, March, 2008, Unpublished

Selinger, E. Postphenomenology: A Critical Companion to Ihde SUNY series in the Philosophy of the Social Sciences series, SUNY Press, 2006.

Social and Ethical Aspects of Biotechnological Practice

Bernhard Irrgang

The globalisation of Biotechnology brings with it not only new economical prospects, but also new risks. According to cautionary principles these risks could be avoided by making efforts towards an essential accompanying Technology Structuring. The following thoughts are not suggestion for complete solutions, but are supposed to open a new horizon of questions for joint interdisciplinary and intercultural project groups.

I. Structuring of Scientific – Technological Innovation

Cultural models criticise particular technological alternatives as inhuman or ecologically harmful and focus on adapted or intelligent solutions. Ideas of naturalness or humanity have always been included in a path-dependent orientation of particular technological development. The substantial paths of individual technology advancement result from an interaction of various selected and limited conditions. With the dynamic of variation and construction, particular fields of technological development, routines of construction and paradigmatic solutions have been worked out. The routines of construction are established in the „State of Technics". In this respect a path – dependency of technological developments results from the practice of technology advancement. There is no central authority which would structure the entire development. Therefore, one has to take various contexts of structuring into consideration, if researchers want to structure a particular frame of conditions.

Technological action is defined as dealing with technological practice which goes along with traditions, paths of development and selective models. Models can be worked out in view of development trends. Here one can speak of trends, but not of preformed technological ideas. Though construction patterns of a technological kind seem to be more than mere social constructions of technological developments. Technology does not arise from one single consistent project, so it cannot be planned in advance, but is being developed out of a gradual constitution process. Yet neither Technology Assessment nor Shaping of Technology can be done by one single project alone but also has to be taken as a gradual process of constitution and reflection. Nevertheless the openness of technological development has nothing to do with irrationality, one can realize it and in the light of this openness one can also act rationally.

Due to the fact that an innovation as a new technological action always includes potential failure, unknown upcoming risks and non-intended effects, there are two ways to handle it: prohibition of every new technology to avoid any possible mistake (tutiorism – taking the safest way) or a permission for new technological actions, provided that the acting person can be made responsible for his innovative technological action, in both a positive and negative case of failure. This implies the demand for Technology Assessment of effects in case of failure of technological action, legal liability and the obligation for research into the risk-potential of innovative actions which means research on safety. Because no particular risk of genetic engineering in cultivation compared with conventional methods has been shown yet, tutiorism in this field of technological action cannot be defended. The question of how to allocate responsibility for innovations is excluded from this. Here the question of structuring technological practise is pending.

The shaping of technological practise is done especially by institutions of technology and regulations with models, economic controlling and through juristic-administrative regulations, it can no longer be understood as a simple technological rationalisation. In the old city-states there have always been social production targets, e. g. the standardisation of bricks. The structuring of technology development is done by 1) institutionalisation 2) standardisation 3) juristic regulations 4) economic utilisation 5) realisation of social needs and ideas of values (cultural dimension). Structuring first of all is based upon the knowledge of technological dealing, orientated on the short-term success. The durability of technological developments results – according to the thesis of knowledge of technological dealing – not from the technological development itself, but from the present cultural frame. Yet one cultural prospect for technological development which would guarantee durability is models, such as the model of sustainable development (Fritz 1995).

Whoever cannot show the harmlessness of a technology, has to expect preventive safety regulations. Yet the recombination of harmless agents has not resulted in any dangerous products. This confirms the fact that a genetically modified organism can no longer be an indication for its possible risks. Genetic engineering has only one risk: You know that a reactor core can melt. But you do not know whether a harmless recombination can turn into a dangerous organism through a line of genetic interactions. From the point of view of evolution uncertain risks and consequences could be no sufficient reason for not introducing a new technology. The obligation of precaution must be seen with different eyes. Because each quantifying weighing of risks and chances is connected with insecurity and especially according to the release of genetically transformed organisms, controversies cannot be excluded, each decision includes arbitrary elements. The attempt of insuring against wrong decisions is done in society for example by law.

Law appears as the right way of structuring and organising technological actions, which legalizes it and as far as that goes, realizes it but also restricts it and points out, that technological action has to be done with responsibility. But this way of structuring should be flexible considering the rapid development. Not least it is a matter of politics, which also has to meet the demands of a global use of genetic engineering. Therefore you can fall back on discussions of structuring society by law or other mechanisms (deregulation). On the other hand it would be desirable to research into the rights of genetic engineering of other countries as well as international regulations and to compare them. It is apparent that national legal systems reach their limits in considering globalisation especially in the area of new technologies. So let us begin with listing the relevant fields of biotechnology development.

II. Application of Biotechnology in medicine and Agriculture

The contribution of biotechnology to the health situation of developing- and take-off countries might be small in the near future. The main cause for diseases in tropical areas can be found in nutrition and in the bad life- and hygiene conditions especially amongst the poor population of developing countries. For research and the fight against tropical diseases, methods of biotechnology and genetic engineering can be used. Companies of poor developing countries or their government are mostly unable to afford those innovations which are essential for genetic research programs (Katz 1995, 44).

Vaccines for particular tropical diseases like Aids, hepatitis and malaria are also relevant for developing countries (Katz 1995, 47). Tropical medical care programs of the World Bank and various world wide organisations must be supported. Besides, industry shows an increasing interest in the market of take-off countries. Its target is the local development of pharmaceuticals. Due to the high costs of genetic engineering, cooperation with industry will be necessary. To make this possible for developing countries themselves, exceptions in patent and sort preservation are being suggested. There is also a demand for Technology Assessment on the effects of on-site technology which consider traditional organisations and the social integration of technology. The betterment of the economic and social situation and the creation of a functioning health system can most probably not be realized by genetic engineering procedures due to low funds. There is a legitimate hope that new vaccines and diagnostic methods for southern diseases will be developed. Because of high investigation costs, the prospect for an economical use for companies which work on the development is only evident in some take-off-countries. Diseases of the poor are especially to be fought in a socio-economic way.

To feed the increasing world population food production will have to be doubled within the next 15 years, yet agricultural floorspace will not be extended. In this case cell- and tissue-technology, which produces virus-free hybrids, can offer a contribution of 15-30 %. Analytical methods to simplify the diagnosis of plant diseases are also of importance. The number of plants, which have successfully been genetically transformed, has exploded during the last few years and already includes the majority of food- and export-plants which are relevant for developing countries. But there is still a long way to go for research to translate laboratory- and greenhouse- results into development of transgenetic types to be used in agriculture.

Most of the plant qualities which are worth being improved like yield, vegetation, duration or fertility are determined as multigenetic. The basic technologies within the field of cell- and tissue-technology of southern plants are already perfected for practise and are routinely being used. Therefore in international agriculture research centres the focus of the work is on illness- and pest-resistance and also on stress-tolerance for improving the combination of substances (Katz 1995, 27). Because of the high state of privatisation of genetic engineering in contrast to the green revolution, problems of technology transfer into the third world arise (Katz 1995, 28). Bio-pesticides have been developed in Cuba, China, India and Brazil. But with Bio-swell, e. g. nitrogen-fixation, they are less successful (Katz 1995, 29).

Genetic engineering prefers countries with a good infrastructure. Latest models try e.g. to connect a permanent protection of tropical rain forests with their cautious and sustainable usage. It is called the use-protection-concept (Katz 1995, 85). Also a possibility for funding of biodiversity could be by connecting it with an offer of processed samples to the chemical-pharmaceutical industry (Katz 1995, 86). Costa Rica, Mexico, Indonesia and Kenya already have exemplary structures for protection and marketing of genetic resources (Katz 1995, 8). Through this the position of at least some developing countries as mere raw material producers was affirmed rather than disproved (Katz 1995, 87): The potential use of large-scale genetic procedures for developing countries is rather low at present. But other biotechnology procedures are already of importance (Katz 1995, 2). Raw materials like cocoa-butter, peanut- and coco-oil which are relatively expensive, could be replaced through biotechnological transformation into cheaper fatty acids, i.e. out of rape-, soy- and palm-oil in the near future. The present local advantages of particular developing countries, e. g. climate and cheaper manpower, could slowly lose their importance. Nevertheless, the introduction of this technology is a question of profitability. As long as these raw materials are cheap, the procedure of genetic engineering will not be accepted.

Foodstuffs often have a high cultural meaning in developing countries, so that in many cases innovations are being rejected. In developing countries genetic engineering and biotechnology offer enzymes, plant-cultures, micro-algae-cultures, bioconservation and procedures for testing the safety and quality of food. Since the beginning of the 20th century 75% of the genetic diversity of useful plants has been lost (Katz 1995, 60). For the protection of genetic resources use is made of the results from biotechnology research. The sustainable usage in each country of their own national plant resources should be intensified, the interests of developing countries should be taken into consideration. Also customer protection in trade with genetic food in non-industrialized countries should be introduced (Katz 1995, 61). Besides questions of production safety and the effects on customers will have to be clarified. Only the Netherlands, Great Britain and the USA have their specific regulations for products that have been produced with genetically transformed organisms, which are unable to reproduce. Here questions have to be clarified concerning the unexpected emergence of new substances, the impairment of bio-availability of a foodstuff which could imply the change of the general potentials of a product (Katz 1995, 133). The risk of uncontrolled export of inferior products from industrial countries into developing countries is realistic. On the other hand there is the risk of import of products from developing countries which do not agree with western safety standards (Katz 1995, 134). The GATT-agreement expressly demands a scientifically approved reason for speculation on safety risks (Katz 1995, 135). Labelling for genetically produced food would be seen as a trade obstacle (Katz 1995, 136). Poor developing countries could be misused as unloading point for products and technologies, which are bad, dangerous and unhealthy.

Also third world countries will solve their problems only if they implement research (Katz 1995, 145). Therefore access is available to biotechnological innovations. Nevertheless, research and technology as well as the entire development should consider national and regional needs (Katz 1995, 147). The task is the fitting of technology into economic, cultural and political frames and capacities of each of the countries. Regarding technology transfer it is advisable to provide only technologies in those areas in which the receiving country could not be economically competitive. Patent- and sort-protection rights should include exceptions for developing countries (Katz 1995, 163), to make technology transfer easier. Technology Assessment has to be established as a complementary method for valuing developing programs. A problem orientated examination perspective should be leading, which also includes the development of both technological and social alternative solutions (Katz 1995, 165).

III. Concluding Remarks

International centres for agriculture research play an important role in the establishment of biotechnology and genetic engineering, e. g. the Rice Research Institute in Manila, established 1962, the Corn- and Wheat-Institute in Mexico-City (set up 1966), the Tropic Agriculture Research in Kali (Columbia 1967) and the Institute for Technology Assessment in Ibadam (Nigeria 1967). All of them have a relative low funding (Katz 1995, 41). They also practise strategies of defensive patenting. This is essential because of the GATT-agreement for protection of mental property. Protection by patent should have a positive influence on development of biotechnology and technology transfer. There must always be a protection by patent for micro organisms and for procedures of biotechnology or genetic engineering, according to these regulations. For plant sorts a protection by patent or any other working protection system or a combination of both should be introduced. The farmer privilege and the breeder privilege within sort protection are still missing. Very broad patents are not useful, therefore their durability is in question. The supporting effect of protection by patent on the development of biotechnology in developing countries is especially doubtful within trade restricted practices. Developing countries should work out a system differing from the patent one and make use of the scope given in the agreement (Katz 1995, 73). It is not reasonable for many developing countries and their farmers to have to pay licence fees for plants, which have been improved on the basis of their own breeding work, without considering this contribution. A recognition of such advance concessions of indigenous communities are not planned yet in the agreement for protection of mental property (Katz 1995, 74).

In theory stress tolerant plants have, particularly for developing countries, a massive change potential in agriculture science. But this results in a move from overused floorspace to untouched areas which again is an ecological disadvantage. According to an improvement of dryness- and salt-tolerance a transfer from useful plants to their natural relatives is not improbable (Katz 1995, 92). Biological safeness in developing countries does not always agree with standards. The risk of release of genetically transformed plants in the third world is usually higher than in industrial countries. Compared with the north we have only little knowledge and experiences with ecosystems in developing countries. Before doing release experiments, we have to ask for the special factors which have to be taken into consideration for the release within developing countries. In addition the safety research on release is insufficient in spite of extensive efforts, especially in China (Katz 1995, 102). On the other hand the problem of spoon-feeding countries according to questions of safety for the protection of their own population should be cleared up and be answered positively (Katz 1995, 103).

Countries in Latin-America especially move towards US-American safety regulations (Katz 1995, 103) and so they inherit a safety problem. From experiences with experiments on small fields they conclude harmlessness or danger of the release on large fields. This is in the least an insecure practice. The substitutes of the synergetic concept demand a case-to-case check up. The Agenda 21 of the Rio-agreement includes an agreement about biological diversity. According to bio-safety records the safety regulations will have to be unified. The development of international Biosafety-Guidelines is essential. Developing countries should be supported with establishment of their own safety-guidelines.

References

Fritz, Peter u.a. 1995: Nachhaltigkeit innaturwissenschaftlicher und sozialwissenschaftlicher Perspektive; Stuttgart.

Katz Ch. et al. 1995: (Hg) TA-Projekt Auswirkungen moderner Biotechnologien auf Entwicklungsländer und Folgen für die künftige Zusammenarbeit zwischen Industrie- und Entwicklungsländern; TAB-Arbeitsbericht Nr. 34; Bonn.

Technological Development and Social Progress

Bernhard Irrgang

I. Introduction

The Industrial Revolution, that took place in Great Britain between 1760 and 1830, began in Belgium, France and Germany about 1800 and about 1810 in the United States. We cannot talk about an industrial revolution in South America, although there are certain forms of industrialization that can be witnessed during 1830. Karl Marx started discussing the question of whether the social problems in Great Britain in the late 18th and early 19th century were caused by the Industrial Revolution or by the population explosion which started in the middle of the 18th century. The reasons for this growing population number can be found in the field of medical and agricultural progress. This question concerns two aspects of technological development and the relationship between the social progress and the living conditions of human-kind on the one hand and the progress of technical development on the other hand. There is no simple answer to this question, since the problem of the interaction between technical development and social progress has proved to be extremely complex, especially in the time of globalization.

The category of social progress is a discovery of historical philosophy during the times of Enlightenment in the middle of the 18th century, which took place shortly before the Industrial Revolution. It caused, by the development of craft, trade and industry an upturn in the economy as a precondition for the fight against poverty, hunger and suppression. All these things are important conditions for the human development of society and social progress. The central factor was first of all the economy, that was still aware of its natural basis among the physiocrats[18], mercantilists (merkantilismus) and within the market economy. This is not true because a working agriculture system represents a precondition for social wealth, especially in the case of a population explosion.

---

18   Physiocrats were a group of French Enlightenment thinkers of the 1760s.

II. Conceptions of the technical progress in the perspective of historical philosophy of the Enlightenment

The philosophy of the Enlightenment thinks about its own conditions of implementation for the first time in the middle of the 18th century. Additionally, it develops a philosophical theory of civilization and cultural development that includes the technical-economic progress. In the middle of the 18th century the first historical-philosophical theories appear in France and Great Britain. They take into account, especially the European one, the technical development within a universal history of mankind. This is a theory of improvement but of nation-degeneration too. Voltaire describes the progress and the history of the Enlightenment as a victory of critical reason over authority and religious belief. It is a victory over prejudices and implies a model which prefers innovation to tradition. The reason should have a critical impartial point of view as a result of which only we can establish a universal history. The aim of the Enlightenment is a universal historical philosophy of the cultural-historical development of societies concerning their economic and technical conditions as well as that of civilization.

Their first representative is Turgot, who propagates an economic theory of human progress including social and technical progress. Adam Ferguson reflects on mainly the anthropological, geographical, ecological and economic conditions for the progress of human-kind. In his opinion the division of labour, the distribution and development of tools as well as the differentiation between trade and industry are to be seen as a driving force with regard to social development. This theory is about the progress of trade and industry and concerns the development of the economic structure of society. These are means for the progress of humanity itself. The financial power of the people and nations are brought together with the principle of utility that is established by David Hume as the fundamental principle of the movement of English Enlightenment. Jeremy Bentham uses this concept and describes the criterion of the greatest happiness for the greatest possible number of people as the basic principle. Adam Smith develops an economic theory of human progress including social and technical progress. Smith gives, „the division of labour and the differentiation of tools as reasons for social and technical development." The central starting-point is the market-economy. Smith sets this economy against mercantilism. The differentiation of technology makes this more professional and the more theoretical way of thinking as well as the construction of technical aids makes it more successful. Smith in his approaches, claims a close connection between technical, economic and social development.

Only Jean Jacques Rousseau fundamentally criticizes the whole process of civilization. He sets the ideal of a natural stage of mankind without any cultural decadence and with natural humanity against this process of civilization. Antoine de Condorcet sees science and technology as the central components of social development. Both produce a moral betterment of society by improving the living conditions of people. Condorcet assumes a close connection between technical and social progress. In the historical philosophy of the Enlightenment, social progress is regarded as a progress of society looking at the growing wealth that produces personal and social freedom and therefore realizes the progress in humanity. This progress means not necessarily higher moral standards among the single members of society, but an additional gain for the public welfare through co-operation. According to Adam Smith the single members are brought together by an „invisible hand". With the help of the division of labour and co-operation in craft, trade and industry a specialization takes place – which is not restricted to the national borders. Smith takes them as guarantees for social wealth coming in useful to all people involved and therefore facilitates the social progress. For those, who cannot participate in the technical & economic progress, there remains no other choice than to trust in the other people and in the welfare state.

III. Theoretical concepts of technological development since the Industrial Revolution

The Enlightenment-conception of a historical philosophy regarding technical-economic and social development applies to a model of technical development that only starts to become technological to some extent. This model established in the time of Enlightenment changes fundamentally as a consequence of the Industrial Revolution and the following transition from technology to industrial technology and technique. To check the validity of this model, we have to briefly reconstruct the terms of technology and technique in the context of their historical developmental. In the general sense, technology is seen as the use of tools, natural processes, gadgets and machines for the realization of certain purposes and aims, which can be achieved through the transformation of natural-processes. Already during the time of founding of big city states like, the Euphrat, Indus and Nile there starts an institutionalization and economization of techniques by producing technical products only for export. We find mass production already in Egypt 3500 years ago, in Greece and Rome 2000 years ago. Specialization, division of labour, professionalization and the more important theoretical aspect of this production-process are developmental factors of the technical development. They lead to an amazing skill base but only become useful to an aristocratic upper level of society.

The historical philosophy of the Enlightenment tries to abolish this privilege concerning the use of technology through the combination of specialization and co-operation, which becomes more and more complex. According to the impetus of Enlightenment – to penetrate the growing complexity scientifically – and apart from the technological sciences the concept of technology develops the technical science of action in the times and context of the Enlightenment-philosophy at the end of the 18th century. It especially considers the production of goods by means of systematization and rationalization of technical procedures. On the basis of this historical development I suggest making a distinction between three variants of meaning of technology:

(1) Technology as a science of technical action and classification of tools and procedures or industries (J. J. Beckmann)
(2) Technology in the sense of social production or of the output in an industrial system (K. Marx) and
(3) Technology in the sense of an interaction between technical & strategic action, especially from the industry- or military-research perspective.

This form of technology is practised especially in nuclear technology, i.e. in information technology and biotechnology. Under the current circumstances it leads to a globalization of technical action as a complex organizing of collective actions. Technology in this sense connects action-in-the-research (research motivation) and technological action in the sense of the definition of Beckmann with economic action within a worldwide operation of globalization.

IV. Models of technological development

Since there are different models of technical development, I would like to present some of the basic models of technical development in this paragraph. This is to make clear, which logical conclusions may be drawn from such a model concerning the question of the relation between technical and social progress. We may distinguish between the following models (Irrgang, 2002a):

(1) Technical determinism: We claim a necessity for the development of technical artefacts and technical procedures that determine the social system. There is a logical development of technical methods causing and dominating certain social forms of development. Technical development in this model is regarded as the main source for economic and social progress (Ellul).

(2) Economic determinism: The driving force of technical development is the economy especially the market economy, not least because of its offer-pressure. Since the Industrial Revolution the market-led forms of production have been the controlling force regarding technical development. The industrial research

determines the technical development and social progress. The money dominates the technical as well as the social development in this model.

(3) Technology as the Destiny: The European philosophy and science follows the principle of reason and causal connection, which drives the technology as a system of the European reason and the European way of thinking and living all over the world. An aesthetic view on the world and a mystic way of thinking could save us from the road to an industrial civilization (M. Heidegger).

(4) The biological viewpoint on technical development: In this concept, mutations are thought of as innovations and selection chooses from the corresponding innovations. The accident of occurrence rules the development, inventions fulfil individual or social needs. There is no possibility of forecast or prognosis in this possible (A. Schumpeter).

(5) Autopoietic interpretation of technical development: This is a theory of feedback. There is a co-evolution between the ecological and the social background and the technical process. Technical systems arise within the bounds of the social context. On the basis of this model certain types of prognosis are possible (N. Luhmann).

(6) Technical dominance over natural, social and human development: Enslaving of nature and man, instrumental rationality dominates all other perspectives. The utilitaristic-positivistic view on reason, that was prepared during the Enlightenment, is predominant. This instrumental reason becomes totalitaristic and terrorizes the human-kind. (Horkheimer, Adorno, Marcuse)

(7) Political planning of technical development: This concept tries to organize and to control technical development by using political or social means, especially with the help of laws and other administrative structures (assessment concerning the consequences of technology).

(8) Evolutorian economy as evolution of institutional structuring regarding technical action based on the path-dependence (Hayek, Nelson).

(9) Contextual Technological Action[19]: Irrgang and Corona in his book „Technik als Geschick? (Technology as Destiny?)", elaborates the model of a technological action (technisches Handeln) within a cultural and social context. This kind of model of action also explores the model of a technological development in our society and can be implemented in engineering sciences and be the basis of ethical acts. First of all, Corona/Irrgang's model investigates the meaning and model of technical action with respect to their development from the cultural and social perspectives. Artefacts like tools, machines or technological structures are the consequences of technical action. They are used for certain purposes and to realise certain goals. This process is defined as the forms of collective technical action and is also oriented with respect to certain forms of technological ethics (Technikethik). This model takes into account the imple-

---

19   Remarks by Arun Kumar Tripathi.

mentation of technical-technological action into the automation and digitisation of technological knowledge in modern technology. (Corona/Irrgang 1999).

V. Technical and social progress

The project of historical philosophy during the times of Enlightenment – is to design a concept of the history of civilization in the sense of a philosophy of development concerning the technical, economic and social progress with regard to natural and ecological conditions of this development. It has lost a lot of its acceptance and plausibility nowadays. There is the widespread opinion, the modern age will be over soon and the age of the Post-History (Gehlen) or the Postmodern Age (Lyotard) has already begun. Moreover, the belief in a deterministic or evolutionary concept of technical and economic development has gotten a lot of supporters. That means it is impossible to organize or even to control the process of technical-economic development. Therefore technical and economic development gets the character of a fate, that is to bear in a more or less fatalistic way, or they give the basis for the legitimacy of an extensive resistance. To an increasing degree the technical and economic competition is regarded as the „struggle for existence" in its military meaning and as characteristic of technical-economic progress. The ethical dimension of humanity is no longer taken into account in an adequate manner.

Economic welfare depends on technical development. It is the basis of a life, where needs can be satisfied and where it is possible to organize ones life autonomously. Those who suffer fundamental deprivation will never be able to spend their life self-determined. The technical-economic development allows only less people world-wide to lead a pleasant life. Therefore the technical-economic development in its current stage is not without good reason accused of being immoral and supporting bad conditions. With such a conception the thesis of Snow about the two different cultures is confirmed. In his concept, the scientific-technical civilization develops rapidly, but the social and moral development – cultivating itself to humanity, moral and culture – lags behind this technical development. The result is a cultural lag and a moral gap. Moral decadence corresponds to the technological-economic progress.

To hold on to this splitting means to capitulate to the philosophical and moral question we are asked by technical-economic development. In many forms we are bound to a cultural and arts-humanistic tradition in the field of philosophy. It is not in every case that this tradition allows us to develop a humanity-concept, which is adequate for the questions of a modern technical civilization. This is obvious especially with questions concerning the medical-technical progress with regard to pollution and ecological problems as well as with questions regarding the generation of energy and molecular genetics. These problems are aggravated by a technical-economic globalization that influences different cultural regions in the world. Therefore progress with regard to the realization of humanity becomes more and more questionable. It means the hope of a historical philosophy in the times of the Enlightenment in a technical-economic progress seems to be dead. Quite frequently the decay and the change in values, caused by the technical development, are highly criticized in many various ways. But even a vehemently presented complaint is not enough.

We need a forced reflection about technical-economic development, we must think in a more rational and enlightened way about development and modernization in the sense of sustainable development. In view of the „Dialectics of the Enlightenment" we do not have to overcome these dialectics, but to take the cultural-philosophical dimension of technical development seriously, in the way it was conceived within the historical philosophy in times of the Enlightenment. We must take into account the intercultural dimension of the technical enculturation concerning all the different regions in the world. To make this possible, we need interdisciplinary and transdisciplinary approaches to deal with questions concerning the technical and economic development. It is important to formulate Husserl's terminology of „Lebenswelt" in a concrete form by taking into account the intercultural aspect and thereby developing a theory of enculturation with regard to technical-economic structures and their ethical assessment. By doing so we have to consider the value of traditions and religious views as well as the world views of those people whose main subject is nature.

Concepts of justice may constitute a central support regarding this question. Not least we have to take into account John Rawls concept of fairness, which connects the concept of justice with the considerations about usefulness and loss. To not let the globalization of the technical-economic development degenerate into a global economic war for resources and markets, we need a philosophical reflection about nature, humanity and society in all continents, based on all the different traditions. The technical-economic civilization and development may be supported by the historical philosophy of Enlightenment. This philosophy cannot be renewed, because we can no longer use the idea of technical-economic progress in such an unselfconscious way like in the 18th century shortly before the Industrial Revolution. But at the same time, it might give us food for thought for a new thinking concerning the forms of modernization between technical-economic development and the development of an idea of humanity as a social and ecological being (Irrgang 1992): The belief in the automatism of the invisible hand is lost. Morally justified conditions are not an automatic result of technical-economic development. Specializations and cooperation under an increasing pressure of competition do not lead themselves to desirable conditions. Therefore the philosophical reflection must ask for institutions involved in the organization of this progress of specialization and co-operation. In view of globalization we will certainly need worldwide organizations and institutions that are able to determine basic conditions for development and to negotiate them on an international level. Mainly they must supervise the things negotiated.

The task of the philosophy is to work out conditions concerning such basic structures of technical and economic development. These tasks must lead to conditions that make life worth living in all the single economic areas that help to realize regionally valid values that facilitate income and ensure the surviving of families in the real world. Worldwide, technical and economic progress alone cannot lead to a social progress, at least not under the conditions of a globalized form of technical-economic dynamism of development. But it would be the utopian views which are to be believed, that the social and humanitarian progress of mankind can be achieved without technical and economic development.

References

Beckmann, Johann Jacob 1777: Anleitung zur Technologie, oder zur Kenntniß der Handwerke, Fabriken und Manufakturen, vornehmlich derer, die mit der Landwirtschaft, Polizei- und Cameralwissenschaft in Verbindung stehen. Neuest Beiträgen zur Kunstgeschichte; Göttingen. (Instructions on technology, or To the knowledge of Trades, Factories and Manufactures, especially those, that concern Agriculture, Police- and Cameral-science. In addition: articles about History of Arts).

Bentham, Jeremy 1988: The Principles of Morals and Legislation (1-1781); New York.

Berger, Günter 1989: Jean Le Rond d'Alembert, Denis Diderot und andere: Enzyklopädie, Eine Auswahl; Frankfurt/M.

Jean Le Rond d'Alembert, Denis Diderot et. al.: Encyclopedia, a study.

Bullinger, Hans-Jörg (Hg.) 1994: Technologiefolgenabschätzung; Stuttgart. (Assessment of the Consequences of Technology).

Condorcet, Marie-Jean-Antoine-Nicolas Caritat, Marquis de 1976: Entwurf einer historischen Darstellung der Fortschritte des menschlichen Geistes, Frankfurt 2-1976. (An outline of a Historical description concerning the Progress of Human Mind).

Corona, Nestor; Irrgang, Bernhard 1999: Technik als Geschick? Geschichtsphilosophie der Technik; Dettelbach. (Technique as the Destiny? Historical Philosophy of Technology).

Ellul, Jaques 1977: Le Système Technicien; Paris.

Ellul, Jacques 1990: La Technique ou l'enjeu du siècle; Paris.

Ferguson, Adam 1988: Versuch über die Geschichte der bürgerlichen Gesellschaft; übersetzt von Hans Medick; Frankfurt/M. (Outline about the History of the Bourgeois).

Gehlen, Arnold 1993: Anthropologische und sozialpsychologische Untersuchungen; 1-1986, 3-1993; Reinbek bei Hamburg. (Anthropological and social-psychological Investigations).

Hayek, F.A. v. 1996: Die Theorie komplexer Phänomene; in: der.: Die Anmaßung von Wissen. Neue Freiburger Studien; Hrg.: Wolfgang Kerber; Tübingen, 281 – 306. (The Theory of Complex Phenomena).

Heidegger, Martin 1967: Die Frage nach der Technik; in Vorträge und Aufsätze, Pfullingen 1967. (Question Concerning Technology).

Heidegger, Martin 1976: Das Ende der Philosophie und die Aufgabe des Denkens; in: ders.: Zur Sache des Denkens; Tübingen 2-1976, 61 – 80. (The End of Philosophy and The Task of Thinking).

Horkheimer, Max 1947: (Theodor W. Adorno) Dialektik der Aufklärung. Philosophische Fragmente; (Amsterdam 1947), Frankfurt/M. 2-1971. (Theodor W. Adorno) (Dialectics of the Enlightenment).

Hume, David 1973: Ein Traktat über die menschliche Natur; übersetzt von Th. Lipps und ediert von R. Brandt; 2 Bde. Hamburg. (A Treatise about human nature).

Husserl, Edmund 1954: Die Krisis der europäischen Wissenschaften und die transzendentale Phänomenologie; ediert von W. Biemel, Husserliana Bd. VI; Den Haag 2-1976. (The crisis of the European Science and the Transcendental Phenomenology).

Irrgang, Bernhard 1992: Christliche Umweltethik. Eine Einführung; UTB 1671 München, Basel. (Christian Environmental Ethics. An introduction).

Irrgang, Bernhard 1994: Gerechtigkeit als Grundlage einer internationalen Umweltpolitik; in: Sozialwissenschaftliche Informationen 23 (1994), I, 40 – 49. (Justice as the basis of an international ecological policy).

Irrgang, Bernhard 1996a: Ein Ethos ökologisch orientierter Humanität als Weltethos; in: Ekkehard Kessler (Hg.): Ökologisches Weltethos im Dialog der Kulturen und Religionen, Darmstadt 1996, 216 – 225. (An ethos about an ecologically orientated humanity as world-ethos).

Irrgang, Bernhard 1996b: Die ethische Dimension des Nachhaltigkeitskonzeptes in der Umweltpolitik; in: Ethica 4 (1996) 3, 245 – 264. (Ethical dimension in the concept of sustainability in the environmental policy).

Irrgang, Bernhard 1997: Forschungsethik, Gentechnik und neue Biotechnolgie. Grundlegung unter besonderer Berücksichtigung von gentechnologischen Projekten an Pflanzen, Tieren und Mikroorganismen. (Ethics of research, molecular genetics and new bio-technology. Foundation with special regard to projects of gene technology concerning plants, animals and micro-organisms).

Irrgang, Bernhard 2001: Technische Kultur.

Irrgang, Bernhard 2002a: Technische Praxis.
Irrgang, Bernhard 2002b: Technische Fortschritt.
Luhmann, Niklas 1986: Ökologische Kommunikation; Opladen. (Ecological communication)
Lyotard, Jean-Francois 1986: Das postmoderne Wissen. Ein Bericht. Hrsg.: Peter Engelmann, Wien, Köln. (The postmodern knowledge. A report).
Marx, Karl 1844: Pariser Manuskripte; in: Texte zu Methode und Praxis Bd. 2, ediert von Günter Hillmann, Reinbek bei Hamburg 1968. (Parisian manuscripts).
Marx, Karl 1969: (Friedrich Engels) Deutsche Ideologie (MEW 3), Berlin. (Friedrich Engels) (German ideology).
Marx, Karl 1974: Das Kapital 3 Bde. (MEW 23 – 25), Berlin. (The Capital)
Nelson, Richard R. 1982: (Sidney G. Winter): An Evolutionary Theory of Economic Change; Cambridge, Mass.
Rawls, John 1975: Eine Theorie der Gerechtigkeit; Frankfurt/M. (Theory of Justice).
Rawls, John 1977: Gerechtigkeit als Fairneß; übersetzt von J. Schulte; Freiburg, München.(Justice as fairness).
Rousseau, Jean-Jacques 1971: Schriften zur Kulturkritik; übersetzt von Kurt Weigand, Hamburg 2-1971, 1-1955. (Papers concerning the Cultural Criticism).
Schumpeter, Joseph A. 1993: Kapitalismus, Sozialismus und Demokratie 7-1993; Tübingen, Basel. (Capitalism, Socialism and Democracy).
Smith, Adam 1978: Der Wohlstand der Nationen. Eine Untersuchung seiner Natur und seiner Ursachen; Hrsg.: H.C. Recktenwald; (1-1789) München. (The Welfare of Nations. An investigation concerning its Nature and Causes).
Snow, C. P. 1967: Die zwei Kulturen. Literarische und naturwissenschaftliche Intelligenz. C. P. Snows These in der Diskussion; Hrsg.: Helmut Kreutzer; München. (The Two Cultures. Literary and Scientific Intelligence. C. P. Snow's thesis under discussion).
Turgot, Anne Robert Jacques 1990: Über die Fortschritte des menschlichen Geistes; Hrsg.: Johannes Rohbeck und Liselotte Steinbrügge; Frankfurt/M. About the progress of Human Mind).

Technology Transfer and Modernization: What Can Philosophers of technology Contribute?

Bernhard Irrgang

Technique- or technology transfer is based in many ways on technological and economical paths, which are often created by European colonization and have been intensified by Industrialization and Globalization. On one hand, the modern age is a constantly developing planetary truth, a truth that labels every society all over the world. On the other hand societies in third world countries have not produced this reality themselves and modernity is an external imposition. This means the modern age turns out to be an unavoidable destiny for them. Traditional modernization and technology transfer is abstract from almost all contextual factors. That is why, technological development and modernization is being compared cross-continentally and assessed with more or less value, without considering the cultural and social contextual circumstances. This supposes on one hand that the western way into the modern age has a model character, is normative and that there are no alternatives to it. Also, it is supposed that the modern age is a desirable objective and that compensation leads to equal final situations.

Requirements for technological standards and for technology transfer are innovations, which constantly promise new developmental paths and a stable institutional setting that can be monitored over a long period. This setting has to be ensured by the cultural system especially by its social-economical dimensions. Also, the dimension of religious world views stabilizes this setting. In Africa and South-east Asia, religion is closely connected to the form and style of life, way of living and to culture. Secularization comparable to the western world only takes place in major cities as they are islands of modernization. For thousands of years, technological innovations and technology transfer have been digested by culturally embedding them and not through modernization. This might vary from place to place but in general it shows the same development. Heteronomy of culture transfers as a balance of deficits meets culturally motivated resistance or is not being noticed. The circumstances of technology transfer are a bit different. It is not being identified with culture transfer and does not automatically lead to a broad modernization but to a form of development with a speed of cultural adjustment – certainly slower than required by modernization. But, this development can mostly be digested with the help of the embedding-paradigm. It is our task to generate forms of modernization with respect for cultural embedding and traditions.

I. Technological Development as Modernization

Globalization in countries with development relies also on modernization in order to achieve a unified standard of technology and civilization culture worldwide. It is supposed that this objective will be reached by natural means and by itself. Hereby, the question of whether that objective model of modernization is reachable at all and which requirements have to be met in advance is not being answered. It would also have to be investigated whether that aim is desirable at all. The modern age is an unfinished project (Habermas, 1988, P. 7). Arnold Gehlen formulated these phenomena into a formula, which is easy to remember: the premises of enlightenment are dead, only its consequences keep on running. Gehlen has separated the social modernization from the cultural modern age. The unstoppable acceleration of social processes appears as the other side (drawback) of an exhausted culture that has merged into a crystal stage (Habermas, 1988, P. 10f.).

Innovative technological development has been a part of modernization since the late 18th century in Europe (Irrgang, 2002b). Today the hopes of enlightenment seem to be realized in Globalization. Techno-structures do not differ from their origin and are not bound historically (Lübbe, 1993, P. 19). This idea is being questioned more and more however. Somebody's identity can be affirmed by a technological universal-culture only in a very insufficient (bad) way. The international techno-culture produces familiarity worldwide but the price is an increasing rate of aging (Lübbe, 1993, P. 26f). Hermann Lübbe following Joachim Ritter regards human sciences as compensation for a natural and technological science without history (Lübbe, 1993, P. 15). This compensation theory is primarily referring to historical cultural sciences that function as a medium, through which culture has compensated for its lack of history by combining monument protection with skyscraper architecture (Lübbe, 1993, pp. 17-19). The compensation theory which goes back to Joachim Ritter considers human sciences as an addition to natural-technological development. The major claim of this theory is the compensation of de-historization by technology and natural sciences. But, this compensation thesis is based on an inherent wishful thinking, which is based on the false fact that the substantiality of past life forms could – imparted by the humanities –regain its liabilities. At the same time the compensation thesis cements the myth of the two cultures. It leads to a way of thinking, which repairs damage after the event.

Modernization was a very popular concept and an ideology in the 50^s and 60^s, in connection with the leave-taking of the old colonial powers and with the anticommunist affect. „Modern age" in the sense of enlightenment describes values such as freedom, individuality, dignity of man, tolerance and reason. The antimodern age as it was first defined in the Romanesque, stands for community, tradition, religion and morality as politics. The industrial societies of the late 19th

and 20th century are marked by technological and economical progress, by growth, functionality, materiality and prosperity. But the progress of a reflexive modern age, which appears at the end of the 20th century, is characterized by realizing somebody's limits, interest in others, preservation of natural and technological innovation, recognition of foreign traditions and co-existence with others. On the other hand, modernization has evolved into colonization, destruction of cultural tradition and forced cultural adjustment (Young, 1995, P. 172).

Technology transfer becomes necessary due to different development paths. Technological levels are closely connected to the thoughts of technological development paths (Irrgang, 2002a). A technological development path is constituted of both technological tradition and technological innovation and it describes a certain final point or stage or a crucial event after a phase of technological development. This is at least a temporal point but often it is connected to the thought of developing at least the technological means. Nevertheless, the speed of innovation often varies and it depends on culture. Technological standards though are not determined only by the technological situation. They are a result of standardization processes and they are a consequence of a successful integration of technologies into current technological practice. Requirements for these standardization processes and successful technology transfer are acceptance and cultural assimilation as well as the combining of technological and cultural paradigms. The necessity of leaders for these processes has to be highlighted at this stage. Co-operation and co-ordination are necessary factors for a paradigm to succeed.

Technological functionality is not decoding in a different cultural way, but the embedding of or dealing with it, without which a machine would not function. Technology by itself without appropriate culture transfers is not sufficient. A secure or ecological technology is a technology within an appropriate context (technology and maintenance). Secure or adjusted technology is a social or cultural status, which is not inherent to technology. Therefore technology has to be designed according to a certain ideal of security, to the user or to environmental compatibility. And these are always influenced by culture indifference to technological functionality, which is often constituted by lthe aws of nature and therefore they are regarded as objective and neutral in value. But handling is a cultural assessment criteria, often formed by prejudice (e.g. through the user) or by somebody's ideas of security and ecological compatibility. These non-admitted prejudices and cultural forming have to be admitted, reflected and discussed. This is a task for a culture of technological reflection (Irrgang, 2002b; Irrgang, 2003b; Irrgang 2007).

According to our western understanding of technology the highest technological standard has the highest degree of automation and rationalization. But for developing- countries the highest technological standard is not desirable because, considering the availability of cheap labour, not much energy or expensive capital should be used. Many subcontractors are not able to meet the high standards of quality so problems with technology transfer occur. Therefore developing countries often do not make use of new technologies, which however do not meet the latest environmental standards. The advantages of labour-intensive and cheap technological means have to be compared with the disadvantages of a higher degree of pollution. It would be best for them to initiate their own developments, which within the development of technological means would enable the criteria of (1) intensity of labor, (2) environmental compatibility, (3) low purchase and production costs, (4) simple handling and (5) profit. The new technologies developed in the west, especially in the area of energy are very cost-intensive. They also have to search for alternatives, which might be found within bio-procedures or renewable raw material.

With industrialization an immense growth of wealth evolved in industrial countries. But more and more the question arises of whether this growth of wealth can indefinitely continue. If a technological standard is to be maintained repairs are necessary and not everything can be completely renewed and instantly be reused as much as it is not possible to make money with the highest technological standard. It is often the use of the highest technological standard, which is also a financial risk and therefore it is not really desirable for countries with financial problems. A requirement for technological standards is innovations, which constantly promise new development paths and stable institutional settings that can be monitored over a long period. This is also a requirement for technology transfer. This setting has to be guaranteed by the cultural system and especially by its socio-economical dimension. Also the dimensions of worldview and the religion of a cultural system have to contribute to this stability.

According to the theory of development paths, innovations lead to islands of modernization. On theses islands procedures of embedding run in a different way than in the countryside because at least a number of modernization and technologization processes have already taken place and the setting for technology transfer and its embedding has begun to be transformed. This means that it is more likely to accept the effects of modernization, but on the other hand the non-simultaneousness of the development speed between those islands of modernization and its surroundings is increasing. Islands of modernization have a bridgehead character for further technological developments and its modernization processes.

Modernization in Europe is often based on criticism of traditions and new modernizations should search for strategies of embedding in developing countries. Instead of announcing a belief in progress without making any differentiations I recommend a sensitizing for issues in the future, which can be practically referred to as a responsibility for future generations. Many of those, who have lost their belief in technological progress and modernization in the western world, still believe in the cultural dimension of western modernization and point out the necessitation of ethics transfer or culture transfer in the area of human rights. Modernization should neither be reduced to a catching-up-industrialization nor to a „catching-up-enlightenment". Considering the meaning of religion in South-East-Asia the claim of embedding modernization in traditions seems to be much more plausible. The reduction of traditional modernization concepts does not aim to abandon modernization but to increase technological competence in the context of an increase in social and cultural competence.

II. Culture and Technology Transfer

Technological development paths have an implicit technological progress and basic differences of standards but they also determine some further possibilities for development. They build an infrastructure and a cross-linked technological structure, whose maintenance has its price. Alternative ways, which deviate from these structures, come with a much higher expenditure than the continuation of current paths. Imitation of innovations from industrial countries where for a long time considered as modernization factors for developing countries. Colonialism supposes the technological and cultural superiority of Europe and the models of development aid and the concept of catching-up-industrialism suppose the difference in standards in technological competences.

The transfer model can have three components (Irrgang, 2006):

(I) Culture transfer i.e. transfer of certain institutions, education institutions, and cultural goods such as science and art have only partially taken place in the colonization process. Cultural ideals, ways of life and the cultural ambience can be transferred. It is not clear yet how much globalization processes will be successful in this area. For a long time culture transfer was based on the ideology of the western cultural superiority.

(II) Technique- or technology transfer took place during and after colonization, though with only limited efforts at the time of colonization. The level of technology of western industrial nations was not being reached except in Japan and South Korea. Technology transfer concerns primarily technologies and forms of economy and is based primarily on values of technological-economical rationality, efficiency and quality.

(III) Moral transfer and maybe even ethics transfer is in general being rejected as Americanization. But this transfer is happening among young people and in urban developing centers (modernization islands) and leads to a broad process of value change. It also leads to estrangement of their self traditions and might lead to the phenomena of uprooting. On the other hand these phenomena can result in positive developments. And estrangement is not an appropriate term because it supposes a relative static conception of both human nature and morality.

Culture transfer is embedded into a new practice not only by transferring or describing ideas or stories. Practices and traditions are a power, which can also exert pressure. Technology transfer takes a technological practice out of its cultural context, in which it has been developed. Cross-linking and embedding processes are being undone and practice contexts lose their meaning and have to be prepared for new cross-linkings, embedding processes and practice contexts. Technology transfer has two aspects, the transfer of economy into the production within the same country but also the transfer of techniques or technologies from the production of a country across its national and continental borders. Sometimes this only needs the exchange of goods which are being analyzed and imitated. Technology transfer in pre-industrial times was relatively easy. What was needed was a movement or immigration of a competent technician who had to lead the embedding process in the foreign culture. Before the 19th century technology transfer was enabled by personal migration and exemplifying. It was bound to the person who preceded the technological transfer. The base for a more intensive technology transfer needs to be built within the system of technological education, which differs from the migration of single technicians. It also seems to support the transfer of cultural goods and values. Today technology transfer is easier than culture transfer and technological practice is a practice led by regulation. At least these rules can be transferred and taught, even if the practice itself is not always transferable. A new practice defines itself only in the continuation and embedding of a development path, which has been developed within the target culture. Therefore not all forms of practice are transferable and the transfer of practices is a problem. Only regulated practices and rituals, ceremonies and production procedures and technological routines are transferable. This means that with a culture transferthe practices of a culture first of all become fixed patterns, which have to progress from this situation and slowly effort has to be put into their embedding.

The inappropriate aspect of technologies, which are being offered to less developed countries, can lie within the characterization of production. It could also lie within the type of products, which are not appropriate for consumption in the target countries. This inappropriate aspect can be addressed if the less developed countries give a more clear definition of their needs and name those factors, which have to be considered in order to design certain technologies and technological products in accordance with their own country. Mechanisms have to be developed, which ensure that technological procedures and products in fact meet the needs and demand of the developing countries. An appropriate politics of technology that considers many questions is possible on various developmental paths. But this politics is set within the conflict of interests of the political and economical elite and within the developing countries itself (Chen 1994, 62). From the philosophical point of view the transferred technologies have to be fitted into the current every day culture both considering structures of production and consumption, even if these will sooner or later be transferred by technology transfer.

III. Technologization and cultural development

A technologization of every day life as in industrial nations has not yet taken place in developing countries. Moreover, the transfer of great technological systems is not an appropriate means for a fundamental reform of the structures of technological production in take-off countries. There are often already competing products in the target countries, which are preferred by most nationally interested consumers. One mistake that has been committed when realizing technology transfer was neglecting every day factors. Every day technologies refer to current needs. In addition there is the dominance of western models. One consequence of this is the necessitation for defining our own models for developing and take-off countries for technological development or to let them evolve from every day live.

Modernization justifies its current revolutionary potential through the blessing of future consequences , just like a futuristic utilitarism. The modern age is not possible without a historical philosophy and maybe not without a utopia of a technological or civilian kind. In opposition Marshall Sahlins talks about the structural power of traditions (Sahlins, 1994, P. 101). Alongside this the structural power of new ideas, institutions and horizons also has to be pointed out. The belief in self-improvements has never received uniformed acceptance. Already Herder has pointed out how the absolutism of our own cultural stage in enlightenment, the so called age or reason or light has consequently led to destruction of other forms of cultures through trading and colonization (Brackert/Wefelmeyer, 1984, pp. 7-12).

In order to create innovations new inventions have to be integrated into the existing network of technological and social solutions. This network approach leads to emphasis on cross-linked technologies and societies. Engineers inscribe social values or norms into a technology and according to these social characteristics the norms and values are being transferred to the user. They then lead to a certain way of acting. Successful criteria depend on mental models, on interpretative flexibility, the effect of demand and criteria for improvements like security, which is an interest of engineers. At this stage cultural barriers within communication between differing engineering cultures and engineering schools can occur.

The task of mediation of self- and foreign interpretation of technological action and technological products has to be changed to a reflexive foreign interpretation. This is the approach of technology hermeneutics (Irrgang, 2001a). It does not only deal with textual interpretation but considers technology with an interpretation of visual and tactile phenomena especially. The cultural interpretation of technological action refers to religious actions and their dramaturgy, because this dramaturgy has influenced technological construction and the development of technological artifacts. Inventions, accumulation, exchange and adjustment are also driving factors of cultural development. The exponentially increasing accumulation rate therefore means that a constant acceleration of the development process takes place. It might appear that growth is irregular. The process of taking over is an even greater source of technology than inventions. It is a process, in which many different inventions from different sources are being combined on a joint cultural basis. The system of all these interrelations builds the organization of culture. Inventions in a cultural area can occur in three ways: As an original discovery, by taking it over from another culture and by adapting inventions from a nearby culture. These adjustments do not happen straight away, but with a certain delay, so we can speak of a cultural phase delay. The cross-linking of cultural territories show a diversified gradation. Adjustment in this sense can be a very difficult process, which requires the construction of completely new social institutions and which can also fail (Ogburn, 1969, pp. 60-67).

Technological modernization in modern times in Europe means experimentally driven technological development. Innovative technological development raises the requirements for the speed with which societies and traditions adjust. This is the case in both in Europe and in Developing countries, where again it encounters other problems not only because the technological development has not met the same standard as in industrial countries. Another problem is the fact that tradition plays a much more important role for cultural development than it does in Europe. This adaptation-pressure on societies and traditions is raised dramatically by technology transfer. Raised technological standards already existed. But today's raising of technological standards has a different effect than it has had before not least due to its cross-linking.

In contrast to modernization philosophies as historical dialectics the modernization theory introduced in this paper should refer to social anthropology and ethnology or ethnography. The program of and hermeneutics of modernization has to be shaped. It is not the aim to compete with ethnology and to become an ethnographer or ethnologist. Philosophical reflection is concerned with finding frame conditions for an extensive theory of cultural modernization.

In Europe the modernization paradigm has three dimensions (Irrgang, 2006):

(I) Modern age and modernization in arts. This is about forms of arts, which are progressive in the sense of futurism and which also have led to postmodern discussions.

(II) Modernization in the area of arts is in fact different to modernization in other areas. In the modern age, the epoch of enlightenment, philosophy is being divided from religion. This refers to the ideological side of modernization.

(III) The third area means modernization through technology in the sense of industrialization or technologization.

Modernization in the form of positivism, profit maximizing, egoism, economics and technocracy is being rejected in most of developing countries, without rejecting the western way of life in general or modernization in total. Overall modernization can be considered as a specific European way of technological-economical and cultural development, which is connected to the idea of experiments. Modernization in the sense of enlightenment includes efforts in the pedagogical area and education of illiterates both in the home country and in colonies. Another idea of enlightenment as a contribution to the modernization theory is the concept of world middle class, which was first realized economically in the world market. But enlightenment is not a world-wide accepted developmental value.

IV. New Concepts of Technological Modernization

Many of the problems that are connected to modernization are neither modern nor neo-modern but some of them are very old. They are connected to technological and social innovations, which have been there for a while and appear now in greater numbers. In this context the new form of modernization cannot be identified with de-acceleration but it is another acceleration of innovations. And this should not only speed up technology development but more and more it has to be applied to social and cultural institutions. The idea of progress as a paradigm of the first modernization was intuitively there even though it was controversial. The definition of a reflexive modernization is neither there in an intuitive way nor can it be understood trans-culturally. Therefore we need another definition for that, which has to be understood as a radicalization of the modern age or a modernization of modernization. Anyhow, the meaning is the cross-linking of embedding and transformation processes of technological and cultural practices. The term, which traditionally was used for adjusting technology to its cultural and social environment, is too simple. It cannot be identified with the dynamic process of today's embedding. The introduction of new technologies in an existing cultural landscape is a transformation of both the introduced technology and the cultural landscape. The phenomena connected with that are much to complex to be described through an evolutionary category such as adjustment. The starting point for the understanding of such embedding phenomena is the theory of path-depending development.

The modern information technologies are for developing countries a possibility for modernization, which at least partially can substitute catching-up-industrialization. And obviously it is more compatible with cultural traditions and the introduced term of technological know-how than the process of industrialization. The idea of technological competence, which gains importance with the new technological revolution, can be combined at this stage with the traditional approach of technology as arts.

However, this phenomenon cannot be described with the theory of reflexive modernization. It is simply the phenomenon of a power of traditions, which influences modernization within various religions in different ways, even if they are on the same continent. Modernization is no project or concept that is independent from previous development paths, which can succeed with the help of rationality. Modernization itself is a radically historical process, which happens worldwide in different ways due to various starting conditions. Unification is not the result of the new globalization wave but in many ways the result is pluralism of circumstances of modernization and therefore also pluralism of modernization paths, in which technology increasingly dominates modernization. Also, this is a crucial point, at which a social science project of a reflexive modernization fails, because it does not give enough consideration to the engine of this new way of modernization, the technological development and technological revolution (Irrgang, 2006).

A demand-orientated technology transfer without the model of a consumer society that opposes many religions and traditions of Africa has to be established. But nowadays the producer-orientated technology transfers are dominant in development aid. A culture transfer or ethics transfer or the transfer of modern world views is not a solution for the worldwide development problems. A new modernization will have to make sure that culture transfer is done in a moderate way and is willing to adjust to each of the established cultures. The regions in Africa and South-east Asia are neither enemies of technology (not even of modern technology) nor are they enemies of market economy. But they are also critical towards global capitalism, as far as this is being considered as egoism.

Modernization should not expect or even claim a capitulation of existing cultures but should aim for a careful transformation of existing cultural traditions. If tradition fights against modernization it can be misinterpreted as a clash of cultures. Such an interpretation though avoids the maybe unavoidable reinterpretation of each cultural, religious and moral tradition in the light of modernization.

In Africa and South-east Asia, religion in many areas still stands for embedding into traditional contexts, which should not be destroyed by rationalistic modernization theories. It is not technology itself, which destroys religion. The western embedding of technology cannot be seen as being a model. Technology and technocratic rationalization processes have to be differentiated. The true problem is modernizing itself as a consequence of European historical philosophy of enlightenment. The west wants, if you allow me to say it this way, a technological modernization on the basis of a modern age and its world view. Also developing countries want this but they have lost hope of keeping up with the technological modern age and its world view. The world society of beings with reason (Vernunftwesen), as it was a wish in the late 18th century of enlightenment, was an ideal without any mistakes but as such it was not possible to realize it (Irrgang, 2006).

Traditional individual enterprises have ecological damages as consequences. So technology transfer should be oriented not towards the needs of western industrial countries, but towards the needs of the developing countries: new information technology, and renewal energy.

V. Suitable technology and cultural identity in developing countries

Technology has no imminent aim. To create wealth and prosperity for everyone, technology has to be culturally embedded and morally orientated. Western technology in developing countries regularly only serves a minority, who uses techniques to reach a higher social and economic status for themselves or their relatives. It is the richer part of the population and especially the one, which is better educated. Therefore the education of the population especially with respect to technical skills and knowledge is one of the most important tasks of governments in developing countries. If they need support, governmental institutions from industrialized nations or NGO's have to assist. Modern Technology can help to increase their own population's health and nutritional situation. But when only a few profit from the newest and highest technology standard and many suffer because of poorness and pollution, then this situation can't be accepted as fair. New innovative technologies must be useful to many members of the population and must be compatible with moral, cultural, religious and social imaginations and values.

Traditional matters of land-, forest-, garden- and ground care, animal protection and support for neighbours should be taken up again and supported in the sense of a comprehensive protection of nature. Modernisation in the sense of European enlightenment, which used to be critical about Christian Religion, shouldn't be an obstacle to the cultural identity of the population. Small, decentralized forms of new, ecologically compatible technologies, which are not too difficult to handle are necessary. In contrast to big technical systems or big plants, which could be built in developing countries only with big financial expense, technologies should be adopted in smaller forms. Especially in biotechnology, information technology and environmental technology there exists decentralized forms of smaller technological units, which could be used by well-educated people even in developing countries. Education especially with respect to science and technology, but also concerning the cultural and religious tradition will always be more important for developing countries, too.

References

Brackert, H., F. Wefelmeyer 1984: Naturplan und Verfallskritik. Zu Begriff und Geschichte der Kultur, Frankfurt.

Chen, E. 1994: (Hg). Technology transfer to developing countries; London New York

Habermas, J. 1988: Der philosophische Diskurs der Moderne; Frankfurt.

Ihde, D. 1993: Philosophy of Technology. An introduction; New York.

Irrgang, B. 2000: Technological Development and social progress; in: Instituto del Filosofia Pontificia Universidad Catolica de Chile; Seminarios de Filosofia 12/13 (1999/2000), 41-52.

Irrgang, B. 2001: Technische Kultur. Instrumentelles Verstehen und technisches Handeln; (Philosophie der Technik Bd. 1) Paderborn [Technological Culture: Instrumental Understanding and Technological Action].

Irrgang, B. 2002a: Technische Praxis. Gestaltungsperspektiven technischer Entwicklung; (Philosophie der Technik Bd. 2) Paderborn [Technological Practice: Design Perspectives of Technological Development].

Irrgang, B. 2002b: Technischer Fortschritt. Legitimitätsprobleme innovativer Technik; (Philosophie der Technik Bd. 3) Paderborn [Technological Progress: Legitimacy Problems of Innovative Technology].

Irrgang, B. 2003a: Technologietransfer transkulturell als Bewegung technischer Kompetenz am Beispiel der spätmittelalterlichen Waffentechnologie; in: Wissenschaftliche Zeitschrift der Technischen Universität Dresden 52 (2003) Heft 5/6, 91-95.

Irrgang, Bernhard 2006: Technologietransfer transkulturell. Komparativ Hermeneutik von Technik in Europa, Indien und China; Peter Lang, Frankfurt am Main [Transcultural Technology Transfer: Comparative Hermeneutics of Technology in Europe, India and China].

Irrgang, Bernhard 2007: Hermeneutische Ethik. Pragmatisch-ethische Orientierung in technologischen Gesellschaften; Wissenschaftliche Buchgesellschaft, Darmstadt [Hermeneutical Ethics: Pragmatical-ethical Orientation in Technological Societies].

Lübbe, H. 1993: Globale Vereinheitlichung durch die Technik und die Vielfalt der Kulturen. Zur Kompensationstheorie der historischen Kulturwissenschaften; in: F. Rapp (Hg.): Neue Ethik der Technik. Philosophische Kontroversen; Wiesbaden, 15-51.

Ogburn, W. 1969: Kultur und sozialer Wandel. Ausgewählte Schriften; ed. von O. D. Duncan; Neuwied, Berlin.

Sahlins, M. 1994: Kultur und praktische Vernunft; übersetzt von B. Luchisi (^{1}1976); Frankfurt.

Young, R. 1995: Colonial desire. Hybridity in theory, culture and race; London, New York.

Ethical Action in Robotics[20]

Bernhard Irrgang

In this paper, emphasis is on the question, whether Computer/Robots can morally act by analysing the difference between the participant perspective (first person perspective) and observer perspective (third person perspective) according to phenomenogical and hermeneutic perspectives. In the third person perspective, Ricoeur's approach to the modelling of acts (actions) without a subjective action is possible. Therefore it seems the Automatic Machines can act on its own without the moral values that are assigned to them. Paul Ricoeur considers the dialectic of a self-identity (Ipse-Identity) and the same- identity (Idem-Identity). In the context of philosophy of language Ricoeur reconstructs the person and identifying designations. It concerns the subject of the act of speech. For robotics and Cyborg, the concept of Ricoeur is interesting and can be applicable to an action without an „acting I" (handelndes Ich). Following Ricoeurean concepts, my paper is an attempt to ask the questions, whether robots can act on their own. Or, the actions of robots can be evaluated?

I. Introduction

In view of the question, whether Computer/Robots can morally act at first the participant perspective (subjective experience: first person perspective) and observer perspective (objective experience: third person perspective) are to be differentiated through the phenomenogical and hermeneutic perspective. In the observer perspective, i.e. in the third person perspective, Ricoeur's approach of modelling of acts (actions) without the subjective action is possible. Therefore it seems the Automatic Machines can act on their own without the moral values that are assigned to them. Paul Ricoeur thinks about the dialectic of a self-identity (Ipse-Identity) and about the same-identity (Idem-Identity) (Ricoeur, 1990, p. 14). Based on the philosophy of René Descartes „I think, therefore, I am" Ricoeur refers to the two aspects of personal identity, whereby this two aspects refer to „I think - I am". This is what, cannot be doubted in accordance with the thinking of Descartes. And, the self executable thoughts in thinking and execution, is a category of existence and a category of thought to be parallel to

---

20   In the paper, to illustrate the translation from German into English language, the translater has used the Italics.

each other. Thus, the starting point is the question of execution of Thoughts (Denken). According to the failure of a materialistic anthropology we can determine human-beings (Menschen) only by its execution of thoughts, which is thereby an execution of „I" <Ich> at the same time. The execution of thoughts can be referred to as an object of studies on the one hand, and on the other hand it can be reflexive. This defines the first tension in the context of the interpretation of execution of thoughts. Second, it refers to the execution of „I" and it reconstructs the execution of „I thinking" in the context of a Hermeneutics „I-Self" and „Same" (Ricoeur, 1990, p. 29). According to Ricoeur, selfhood implies otherness to such an extent that selfhood and otherness cannot be separated. The self implies a relationship between the same and the other. This dialectic of the self and other contradicts Descartes' cogito („I think, therefore I am"), which posits a subject in the first person (an „I," or an ego) without reference to an other. Thus, the hermeneutics of the self differs from the philosophy of the cogito.

In the context of philosophy of language, Ricoeur reconstructs the person and the identifying designations (Ricoeur, 1990, p. 39). It concerns the subject of the act of speech (Ricoeur, 1990, p. 60). For robotics, the concept of Ricoeur is interesting and can be applicable to an action without an „acting I" (handelndes Ich). (Ricoeur, 1990, p. 73). Following Ricoeurean concepts, one must ask the question, whether robots can act on their own. Or the actions of robots can be evaluated? In the context of possible behaviours of a robot, it probably does not deny that generally, robots can completely fulfill a pattern of action (scheme of action). In the most general case, an action pattern (action-scheme) is present, if at least the intentionality and the goal of the action are preliminary modelled.

The <Self> and <I> are to be differentiated, since the self description is the category of reflection. „The Self (Ipse)" and „The Same (Idem)" are also to differentiate as the actual description for an Identity. Philosophy of subject or philosophy of „cogito = sum" are to substituted/replaced with the hermeneutics of self. In the philosophy of history, the phases of overestimation and underestimation of the subject are followed and studied. In a hermeneutics of the self, it is concerned with the determination of self i.e. the subject of spoken, action, narration and of moral attribute. Selfness and subjectivity can be understood as a dimension of human action and the unity of self. Self testifying and attributes are here by inter-linked by others. The person is to be understood as a fundamental single thing (individual).

Now, are there human actions in Computer Science or in Robotics? The computer extends the range of human action (activity of human-being). First, programming is to be defined as an action within computer science. Computer science offers special teaching of dealing with the complexity of problems. This transition, however in practice does not lead to the desired objective of the explanation, but it becomes a pure action in the sense of the manufacturing process

(Kastendiek, 2003, pp. 157-159). Programming can also be understood as a medium to provide the knowledge and technology (mediation of knowledge and technology). Programming is a practical activity, which is developed in computer science (Informatik). It is based on the mediation of theoretical and mathematical aspects of history of computer science, which in turn are based on the practical solution of human and social problems. Thus it is not a scientific method, but it can be defined as an action (Kastendiek 2003, pp. 161-163). But their authors attribute the character of action to computers themselves.

This is inclined to attribute functionalistic & operational parameters of responsibility to the computers. Computers are made responsible for their decisions and thereby it demands that we make them more trustworthy as this has already been done by an American General some decades ago, then indeed it does not violate the Taboos: Because, computers are neither moral nor social beings, although their common employment can produce extensive social effects. Today even more decisions in society are left to the individual information systems, because human-beings are not able to accomplish the task of making a responsible decision due to extremely short response times. Therefore these decisions are no longer made by the individual human-being. The individual human-beings are not responsible for such decisions due to short response times. Klaus Haefner realizes the first step in the actual inevitableness of the transmission of responsibility attributed to computers, which is an integration of human-beings into a very complex overall system. It is increasingly responsible for human-beings. Indeed, one must differentiate between the descriptive analysis of causality on the one hand and the normative expression of responsibility on the other hand, in the sense of the someone-real-responsible-person to be made (jemanden-wirklich-verantwortlich-machen) (Lenk, 1990, pp.104-106).

William Bechtel and John Snapper are working on this concept. They specify quite reliable conditions for the flexible program-controlled decision systems, which according to their opinion permit them to provide legal and moral responsibilities to these systems. So far, it has been problematic whether the operator, and owner or the manufacturer, programmer, software producer etc. are responsible for the computer errors (Lenk, 1990, p. 107). Who is going to take the responsibility for such errors in the system? Like Snapper and also Bechtel he uses the Concept of Responsibility term only in the sense of descriptive-attribute to define the problem and give solutions. The normative expression in the sense of someone-responsible or someone- -holding responsibilities is not discussed (Lenk, 1990, p. 111).

This is the underlying mission: In the long run, the new systems are the robots or more highly networked systems, have the characteristics to be able to „act" more autonomously, to be able to unreel without being able to reach „autonomy" in an ethical sense. To unreel the behaviour programs in the long run certain framework decisions have to be met, which may have serious consequences for human-beings (humans). They could lose their characters of tools and become autonomously active participants. In the robots existing today, humans must still intervene controlling, steering, supervising, but the degree of automization is growing. Thus, it raises the responsibility question for the behaviour of robots and AI expert systems, their technical designers and users. The early discussion about the dispute with computers, was limited to the topic of artificial intelligence (AI), but today's debate under the spotlight of concepts is constrained to such areas as data processing or networking. Thus, the computer can be seen as a tool, but also as an autonomous machine. The subject of action decides thus, it wants to see which characteristics lies in the computer. Characteristics in the scope of possibility is to be registered regarding the level of action with acting computers. The computer can be regarded as an all-purpose tool or a universal tool (Kastendiek, 2003, pp. 9-11).

In addition, the thought (Gedanke) of a machine-similar action should be introduced. First, we have to differentiate between action and behaviour. What similar machines make as actions, is to be seen not simply for humans. It needs a lot of training and substantial efforts of will for humans to implement actions in this way, for instance military drill. The desire of human behaviour must be for a long time only learned in the sense of a machine-similar drill practice. The thoughtless routine must be internalized and constantly repeated. Afterwards, the machine actions and human actions are determined. Nevertheless, actions which do not follow explicit rules, may still be summed-up (subsummierbar) afterwards under a rule. Retrospectively formulated rules do not apply however to the future. Whereby one can split and divide behaviour? That is one of the central problems of methodical kind in the description of documents and behaviour (Collins, 1991, pp. 30-45).

Thereby, it gives a clear picture of an increasing authority and competence of machines. If humans, for example feel too tired to implement tasks of routine, they could delegate this whole process to their machines. It concerns the extent to which humans in the situation delegate their actions to machines. We cannot delegate actions, rather we can only delegate the behavioural co-ordinations of the act. We can delegate the action successfully only up to this degree, which is the part of an action and in sufficient measures is suitable to be reduced to a specific behaviour, that is recordable without the loss of a formula. We can say, the dancing steps coincide with the steps of many participants, if we describe a physical activity of the production of standardized artifacts or mental activities like those of counting (Collins, 1991, pp. 65-71).

II. Perspectives of Ethical Acts

Computers are part of our culture. We have expectations that those computers think like machines. In addition, we have been inclined to attribute intelligence to all which was invented in recent time. The machine-machine interface however is decisive. In the Turingtest, it concerns the man-machine interface. The test takes only 20 seconds and that is not sufficient time for a genuine evaluation of machines. Machines with an experimenter can be trained (Collins,, 1991, pp. 187-194). A programmer can never anticipate what an interrogator may say or ever ask (Collins 1991, p. 203). The power of a computer can only be understood in the context of the social groups, to which it is assigned. Human beings compensate with the defects from artifacts. What can cause or additionally teach and carry out an expert system to a social group? Nothing can be completely described. Then machines can take over the competent-oriented jobs of workers, who then takes over the work for the incompetent? Well, this can apply, but this applies alone to behaviour-specific competence. The only error is to believe that a tool is an actuator (Akteur) i.e. an action. There is a substantial difference between the wayhumans act and the way in which machines copy this document (action). Intelligent machines are however the most useful and interesting tools (Collins, 1991, pp. 215-224).

It is to be considered, that robots have the shape of humans and are able to be similar to humans, but not in the near, their abilities however can be similar to humans. They need the correct environmental model, thus they are not environmentally open (friendly) like humans. They need a constant control and maintenance and they need stable environment conditions in order to do the correct behaviour for the day. Probably, the instinct-secured behaviour of some animals is more autonomous and more similar to humans than talented robots in the foreseeable future. So far automats have only had very limited functionality. They lie within a range of pre-rational intelligence, within which certain human competence can be simulated. The robots of the future will be copied even further from their models, humans or animals. And, in robots the artificial and biological elements will unite. Computers, especially interlaced computers are disembodied, artificial Intelligence. The realization of this intelligence in a body-free Cyberworld (sort of disembodied world) is one possibility and the realization in robots is another. We will have to wait and see, finally which way forward in AI development is the best.

So that artificial intelligence will become intelligent, it must have an artificial soul and it must not be limited to a pure mind and pure cognition. In addition, the insight is necessary that feeling and motivation contribute to the increase of intelligence actually to a considerable degree and that they in the computer are possible (Doerner/Spitzer, 2002, p. 24f). Doerner understands the imbedding of the human action regulation, and also the imbedding of the cognitive processes by feelings into a structure about modulation and action tendencies, which are for their part situation-oriented. For the fact it is important that the brain puts on memory minutes sequentially for our activities. With the memory however, I have the basis for self modification. The free will is the released will or the released motive. Man will build artificial souls, because we are making an attempt to copy the psychological processes to make us understand ourselves much better. Human beings should know substantially more than is so far the case (Doerner/Spitzer, 2002, pp. 33-36). Whether however a human soul could be designed independently of a human body in another form other than a most rough model of a soul, I hold this idea to be very doubtful. Besides, it would have to be clarified, which psychological characteristics are to be given to such a soul.

The options of replacing humans can represent a specific perspective in robotics. It is objected that the communicative competence of humans is not in principle technically replaceable. Technology proceeds already from the outset, where it substitutes human activities, procedurally differently to nature (Decker, 1999, p. 23). Cognitive robotics is for engineers, at the beginning of a technical conversion of only really new body-soul theory, namely the functionalism (Decker, 1999, p. 37). But, it ignores the embodiment of humans, which is not able to be simulated. In robotics, increasingly it goes around functional coupling between nerve cells and technical systems in the form of neuro silicon couplings into cell cultures (Decker, 1999, p. 41). In order to be able to take responsibility, technical systems for humans must be controllable. By a complementary system of organization work quality in the double sense of competence-promoting tasks and effective achievement contribution becomes possible. Rarely, accidents occur and one must consult the collectiveness of beneficiaries, in order to compensate for the damage when the technology malfunctioned (Decker, 1999, p. 65). This refers both to production and consumption.

Many repercussions in the 20th century take place against the privileged position of humans in 20th century. Now machines appeared as a concurrent for the privileged position of certain humans. Humans are not any more similar than machines. Computers are today in symbolic algebra better than humans. Computers with the correct software really think about facts, make decisions and have goals. It is the goal to build robots which act, as if they seem to have fear or simulate the fear. We draw a limit of comparison with our feelings. In the animal realm, we are still safe in our position, therefore we confess some feelings towards animals (Brooks, 2002, pp. 172-189). We find ourselves already using the word machine for humans. The machine operates in the same way as humans. John Searle subordinated, human brains and nerve cells somewhat particularly, which bring out the consciousness, but he does not give a reference to it anywhere. He also does not say why a system on the basis of silicon cannot have consciousness. Artificial life (AL) and artificial evolution make available whole new thought categories. The presence of models of neural networks has still to be developed. The question about consciousness is a very difficult question. Robots as our slaves are the beginning of morally justified slavery. But, if robots have the status of humans, then also this form of slavery has to be stopped (Brooks, 2002, pp. 192-215).

In 20 years our personal computers will be 1000 times more efficient than today and will then exceed our intellectual capacity. This is the forecast of Rodney Brooks. Perhaps we can build a computer, which is more intelligent than we are, perhaps not. Do robots then represent a danger for humans? Is it in the paradigm of condemnation or release by robots to be arranged? The human resemblance of robots could be limited. Some basic assumptions of the robotics should be discussed, in order to tighten their potential limits: (1) the question of their independent reproduction: Machines, which repair themselves and reproduce themselves, would however depend further on an infrastructure. Humans are worried about these developments. The supply of materials and also the maintenance of production of computers, are the task of humans. This infrastructure is still much too complicated for robots. Also in the chip production in next 25 years much human participation is still necessary. (2) robots do not have feelings, but sentimental robots will be designed. (3) the survival instinct of robots; Robots, which guarantee their own power supply, are not conceivable in the foreseeable future, because robots in the long term, will be dependent mainly on electricity, which is operated by humans. (4) Loss of control of humans; Such a development does not occur overnight and not at a rapid speed. We will notice it, if it is imminently possible to build machines which we no longer can control.

The three disappointments of human narcissism, Copernicus, Darwin, and Freud, were the milestones on the way to perceptively and finiteness of the human mind and its embodiment (bodily existence). AI could be in the next stage and teach us to read a new interpretation of hermeneutics of anthropology. A condition for this, however is that we learn from the visions, for example, to learn from the discovery of Hans Moravec's illusions. In billions of years of an untiring arms race our genes succeeded finally, in booting itself out. We prepare a post-biological, supernatural future for us. In this future world, the new human kind will be torn-away by an eagerness for future changes and displaced by its own artificial descendants. Today our machines are still simple creatures, which like all newborn babies and children require, the parental care and welfare service and are hardly intelligent enough to be designated. Since these children of our spirit are not dependent on the coming of a stagnant course of biological evolution, they will unfold and be unrestrained. Thus, it comes to an awakening redemption between the intelligences. Since the industrial revolution, it has become possible to replace the human bodily functions by machines. We are now very close, to realizing the point at which practically each important physical and mental function of humans will have their artificial counterpart. The primitive evolution began with reproduction of clay/tone crystals. Our genes and body from flesh and blood will soon lose its meaning. It is not difficult to imagine human thinking, free from the connection to a mortal body. Hans Moravec pleads in such a way for a post-biological world (Moravec, 1990, pp. 9-14).

III. Autonomous Robots

Robots, which can act „autonomously", or to be more exact, which can behave in many situations perhaps similarly to animals, remain technical products. They remain tools and do not become doer (acting) subjects. How should they become subject? (How can one attribute subject to the machines?) Only by a miracle! Robot-action belonged to the cases of actions without a doer (acting) subject. Since the industrial revolution, an illustration of technical action patterns in machines is to be seen. Here it develops with the reproduction of the spin procedure in machines and with the linkage of weaving looms first integrated machines an illustration of more complex human technical action. A modelling and simulation of human competence inclusive application tools is therefore not new in principle. Indeed it could be possible by robots in the future. It is an extension of the basic pattern based on non-technical courses of action such as perception, mobility etc. It can also be used for non-technical purposes. Thus a further mechanization of everyday life will take place. But in principle, this is also not new.

A robot does not mutate to the acting (handelnden) subject. The acting subject with a person or a personality is not the product of a genetic code or a computer program, even if many reductionistic science programs believe in this project, and no system theory wants us to believe this. Now it is in principle conceivable, that robots will be further developed and finally also with their own energy balance be at least partly equipped and possibly be able, to act universally, is multifunctional. Also, for the completion of several tasks to be programmed, so that by the interaction of everything of authority it possibly appears it in such a way to some humans, whether an autonomously acting subject would act here. The question is however actually open, with programming like that in what respect something is at all present, like a consciousness of their own courses of action in the machine. We may not subordinate these simply, because the action forms of such a computer as to simulate human behaviour in their aspect of third person perspective. The aspect of execution of a human agent or a human action pattern cannot be simulated in a machine, because the execution perspective concerns the first person perspective. To that extent possibly a comparability is here, given in the external aspect of human behaviour and in the robot behaviour by a similarity of courses of action.

Of course we can attribute a status to an animal or a robot exactly in the same manner as humans award a certain status to others in a certain role. If we award dignity however in an almost indiscriminate way, in the moral sense to robots, animals and humans, then the concept of dignity in a dramatic way devalues itself. It cannot be consulted then around a certain privileged position of humans. If we do not want to maintain this kind of difference any longer however, then in principle the totality regime of torture methods is finally justified. To that extent, it appears as necessary not to follow the efforts, to aim at a condition of posthuman. But to work out for a suitable anthropology, which can be established on the basis of a humanity understanding, which however is not against the modern realizations for animals and computer action patterns, i.e. to maintain the cognitive structures. I assume that the simulation of cognitive structures does not fit by any means automatically with knowledge, just as few as the simulation from actions to humans. It is a morally attributable cause of (zurechenbaren) action. Without this in a strict sense to be proved; It gives a series of pragmatically forms of evidence, why a human action and morality remains embodied and is mediated or remains intact in an embodied mind. To that extent, probably an isolated brain cannot act morally in a technical matrix. Here, however a line of limitation is present, on which one must contemplate.

A concrete observation of „autonomous" robots concerns actual potential of danger, which can arise from them. Possibly household robots do not give consideration to babies, which come in their way. Household robots displace the babies even actively to another place, whereby this treatment of household robots is not good for these babies. Also, different accidents, e.g. on provision courses of household robots, are possibly not an assessable risk. On the other hand, it could result in that we in fact design and build a human-like household robot that is good for babies. There are possibly, other special technical solutions for special uses, through which robots not like humans can go shopping. On the other hand, there are efforts to build a universally acting robot which could become indeed human-like. But why should we actually build such universal robots? At least from today's perspective it appears as meaningful to design robots, computers or expert systems which can fulfill certain specialized tasks and also we can finally make use of them. In such specialized function robots are the tools, which remain dependent on humans in a certain way and so that they belong to the area of responsibility of humans, similar to the traditional technological application.

Why could it be meaningful to build a universally applicable robot if there are humans, who can fulfill such tasks? Perhaps in space, where robots can work better together with other robots, because in space human life is endangered. One will have to contemplate such areas of application, and need a knowledge to work out the hazard potential from limited working robots within the specialized ranges. Afterwards estimations and considerations must be made as to whether we finally want to design a universally applicable robot and so possibly generate a new technology, which will dismiss the human control. From an ethical perspective a technology, which could completely turn out of control and cause heavy damage, may not be built, Because, this is the limit of a potential responsiblilty of assumption of capacity of each human, which can also be valid for collective responsibility of humans.

In all other respects, the question still exists, why artificial humans and robots must be absolutely human-like, so that they are useful. I assume that it is the fascinating dream of humans (Faszinosum Mensch), which energizes us to convert such thought experiments into practice, in order to find the technical models of ourselves, which will give us an understanding of who we finally are. But we can get success on this by technical artifacts and technical models. I also assume that such universal and unspecified robots with large flexibility in their behaviour patterns and a due portion of learning competence, as well as a certain human likeness, nevertheless do not finally become humans, which are after all equipped with knowledge in a theoretically philosophical sense or with action competence in a moral sense. Even if they are possibly a useful simulation for the everyday life of a human being. It must be considered that, potential of humans is more comprehensive than that of a common-sensical understanding of the technologized everyday life. Also, possibly if robots' action patterns as quasi autonomous process cycles are programmed and independently it can learn in certain way, with the fact that by no means these robots can also reach a consciousness of their actions and execution of their action patterns. We do not subordinate consciousness to a computer machine in order to examine content wise the arithmetic operations. Also a text processing system probably does not understand anything about the texts that it works.

But even if the complexity of programming extremely increases and machines can learn it how should it become possible that an arithmetic performance capacity, which obviously produces no consciousness can develop a consciousness or even self-confidence? Perhaps we can model t humans or with the same width unit humans quasi as a programmable blueprint to sketch for a robot. But actually let humans themselves be simulated as not a determined animal in a program and can this simulation be regarded as the original one? Liberty or freedom can be simulated in the random number generator (Zufallsgenerator), but human liberty or freedom can't be simulated. Then it is obvious for humans to deny liberty or freedom. The model or the copy even of an object is from the same material, and has the same form. We nevertheless, usually estimate and asses the original clearly higher. But as copied paintings and the original paintings have feature in common, robots due to their material difference and organizational difference with humans will not be able to reach the proximity.

IV. Decision Making

In a thought experiment (Gedankenexperiment) one can equip humans gradually with more Exoprothesen. Thus in such a way developing prosthesis structures, after a complete replacement of all organs become Androids (Erlach, 2000, 36). Here Erlach err, if human courses are supplemented by the f prostheses and, replacement with artificial organs and cosmetic surgery, then it does not become an artificial human, Android or a robot. But it will die, because sometime in the overall system humans will break down. The thought experiment leads to an interesting result that there is theoretically a far transition field between humans and robot, what is relevant however is the question, to what extent this process is technically feasible. The success of this process does not depend only on the technological fate, but particularly on the maintenance of a vital function of humans, with the replacement of parts. This limits the thought experiment in an attempt of its execution (Durchführung).

Without acting I (handelndes Ich), the goal structure of action pattern can also be ethically evaluated. If a goal structure can be reconstructed, an action is present, also if no acting I (handelndes Ich) is to be identified. I can judge whether a robot carries out a morally positive or not-positive action pattern. However, a non-embodied (disembodied) (non-human) action is only in the abstract sense an action, for whose consequences are also attached to the robot and held responsible. But, possibly the programmer, who sketched the action pattern and implemented it in robotactio. Action is itself an interpretation construct (construction of interpretation), defined a write-up phenomenon. As a pure physical procedure does not differ from the regarded actions of events. Therefore, positivists come to a conclusion also on the idea, that these do not provide the moral actions and human liberty (freedom). But our self experience reflects otherwise (Irrgang, 2005).

The psychological & thoughtful (denkerische) identity, based on the continuous execution of thoughts (Denken) (materially spoken, non-stoppable brain processes), constitutes that the embodied execution of the beginning of life of the brain up to the death of the brain. In order to be morally responsible however, an act needs a participant, who is characterized by personality or subjectivity. Robots can proceed according to a certain action pattern (goal structure and consequence of evaluation), which can also be evaluated morally. The robot cannot be brought to account, since it was not the actual participant, but it was a generator, that is a programmer of the action patterns. What can now identity and selfness of an embodied execution mean? The crucial and decisive dimension of reply is based on the fact that the wholeness of an execution remains protected. The wholeness of an execution however can be judged in the long run only in a retrospect by its conclusion and after it is completely evaluated. This wholeness of an embodied execution begins with humans sometime between conceptual design (Konzeption) and birth and ends with the death of the brain. With a robot, the technical execution begins as a process without participation of subjectivity or personality by others. Engineers take a robot in an enterprise and sets him to carry out an operation; in contrast to the embodied execution, the technical enterprise of a robot can also be interrupted.

V. Conclusion

Robots do not become humans and cloned humans or genetically redesigned humans will remain humans. To that extent, a certain naturalness of human embodiment (physicality) is a central issue to establish an anthropology of homo faber in the 21st century. The difference between humans and robots is not to be removed. A Cyborg is defined as a cybernetic organism and is finally a robot or artificial human. This concept shows most differences, perhaps even opposites. Humans with an artificial heart remain humans, the same with heart tissue. Humans, also with a genetically improved code, cannot mutate and become a robot. Thus, the difference between „natural" and „artificial" is preserved within the area of anthropology despite all sharpening (Verschleifung) in a certain way. The Kaleidoscope resolves the clear differences between naturally produced humans and technically designed robots, but it does not remove it completely.

The human beings who are reconstructed almost totally after a fatal accident and equipped with prosthese, do not become a Robot: he/she remains, a human being when he/she survives. Humans can be supplemented prosthetically, but he/she can't survive his/her brain death, based on our current knowledge. One could use parts of a corpse (dead bodies) theoretically for the production of robots. That, this allow them to remind themselves of events from the life of a dead one, is almost improbable. The designer baby likewise becomes no Cyborg in the sense of a robot, even if the most general definition of Cyborgs permits this interpretation. I would thus, suggest to clearly differentiate the concept of Cyborgs from that of a Robot. Nevertheless, we can make the reverse thought Experiment (Gedankenexperiment) and try to build computers into humans! This will not be technically possible, because probably a brain is not able to control a robot. Of course one can be able to build and transplant some chips. The limitation between humans and robots, perhaps falls into the category of fantasy, not however within the range of the technically realizable one. With their limitations between morally attributable (zurechenbaren) acts (first person perspective) and machine regulated, implemented behavioural patterns which are capable of being implemented (third person perspective). It is an amazing fact, which fairy-tales researchers are ready to believe, if it concerns building artificial humans.

The identification of a robot with a natural-born cyborg has disastrous consequences: Either genetically designed humans as well robots do not have human dignity, or human rights must be awarded to robots. And all this can only be accomplished, due to a weakness of concepts (Begriffsschwäche) and thinking, the concepts of reflection incompetence were used inaccurately. Also the thought that between genetically adapted humans, who are wrongly called Cyborgs and robots, who are also called Cyborgs, in them some kind of sex or even love could be possible, which neverless strains our fantasy, to a great extent. Here, the film „Blade Runner" conveys the correct answers in the form of literary mediation. In addition, technologically reprogrammed humans become the designer human being and humans with tissues technology produced for organs for a Man-Machine hybrids or natural born Cyborgs, but it cannot be robots.

The difference between humans and robots remains at least so far clear, as humans are propelled to be nourished robots by technical (in particular electrical) energy. Extreme forms of man-machine hybrids or human brains in a nutritive solution i.e. in a technical matrix remain for technical like ethical reasons science fiction. A brain in a nutritive solution or a bat i.e. in any matrix would actually be posthuman. It would be only in the most reduced way a spirit, because this brain was at least bound to a body. Therefore in a rudimentary form there probably still remains parts of a body consciousness, which must now however in the nutritive solution from a human body separated „live". This brain is the conceivably most extreme form of a Cyborg, however this does not-not become the robot. Probably such a brain also still has memories of the embodiment (physicality) and a subjectivity developed in former times. A conception of subjectivity, which may develop a brain in a nutritive solution, is to be apostrophieren from the perspective of the human brain today living as an embodied mind as strange and posthuman. A human brain in a nutritive solution or a matrix is not human, but remains a brain in a nutritive solution. Only if such a brain were able to control a machine the question about the morality of the courses of motion i.e. actions of this hybrid thing arises. It is done in such a way, when whether any replacement (ersetzbarkeit) of physical (corporeal) functions would be possible with man-machine hybrids by technical surrogation (Surrogate). This corresponds to the logic of technical expansion, but probably not the logic of human embodiment (Bodily existence) (Irrgang, 2005).

The Science-fiction of the future of humans usually comprises of negative utopias of the posthuman. Posthumanity is growing on the basis of a wrong picture of humans and a wrong view about technology. On the other hand only philosophical reflection helps us to understand the clarification. The negative utopias proceed from the failure of humans, but the Humanism tradition wins in technical thinking of strength. Therefore, the discussion over posthumanity is probably not much more than the last rearing up of an ending cartesian dualism. Homo faber artefactus, the autonomously acting robot, is not by any means the reality, not even a silver-lining on the horizon. But it casts its own shadows (shade) ahead. It is the time, to speculate about the century, over a ostensibly and supposedly posthuman human-being (Bodily existence; Menschsein) in its way of technical existence and its Technosphere about an anthropology of technology in the 21st century, that advances with technologization of humans and its subfunctions with an unforeseen speed.

References

Brooks, R. (2002) Menschenmaschinen. Wie uns die Zukunftstechnologien neu erschaffen; übersetzt aus dem Amerikanischen von A. Simon. Frankfurt.
    Collins, Harry M. (1991) Artificial Experts. Social Knowledge and Intelligent Machines. Cambridge Mass./London (1990).
    Decker, Michael (1999) (Ed.) Robotik. Einführung in eine interdisziplinäre Diskussion Bad Neuenahr Ahrweiler.
    Dörner, Dietrich, Manfred Spitzer (2002) Vernunft – Gehirn – Computer: Was bleibt vom Menschen? Bamberg
    Erlach, K. (2000) Das Technotop. Die technologische Konstruktion der Wirklichkeit Münster Hamburg London.
    Irrgang B. (2005) Posthumanes Menschsein? Künstliche Intelligenz, Cyberspace, Roboter, Cyborgs und Designer-Menschen - Anthropologie des künstlichen Menschen im 21. Jahrhundert Steiner Verlag.
    Kastendiek, Antonia (2003) Computer und Ethik statt Computerethik Münster.
    Lenk, Hans (1990) Information zwischen Verantwortung und humaner Eigentätigkeit. Chancen und Probleme der neuen Informationstechnologien für Gesellschaft, Moral und Ausbildung; in: B. Irrgang, J. Klawitter (Hg.): Künstliche Intelligenz; Stuttgart, 99-115.
    Ricoeur, Paul. (1990) Soi-meme comme un autre Paris (Ricoeur, Paul. Oneself as Another. Translated by Kathleen Blamey. Chicago: The University of Chicago Press, 1992.).

Note

The earlier version of this paper is published in the Conference Proceedings of the Sixth International Conference of Computer Ethics: Philosophical Inquiry (CEPE2005, July 17-19, 2005, Enschede, The Netherlands) eds. Philip Brey, Frances Grodzinsky and Lucas Introna, CEPTES, Enschede, pp. 241-250.'

Epistemology of Biotechnology

Bernhard Irrgang

In order to get an epistemic understanding of genetic engineering the theoretical scientific concept of laboratory science is being confronted with the technology science concept of implicit knowledge. It turned out that with the stage of synthetic biology as a contrastive technology of also new live forms technological element gets the upper hand within laboratory praxis. The number of technological procedures and technological modelling next to experimental methods is increasing. This model is being accomplished by the creation of complete synthetic life forms, the development of renewable resources and new material.

Genetic engineering like nuclear technology and information technology is believed to be an example of technological research, which sells itself to industry and produces incalculable risks. Technological research is usually regarded as the continuation of instrumental-experimental empirical natural sciences. Theoretical sciences and epistemological concepts are largely missing in this area. One does not need scientific theory in order to operate science, especially not with technological research. But the lack of methodical self reflection of a scientific practice provokes insufficient design concepts.

On the other hand, the traditional scientific theory of technical sciences for modern biotechnology does not help too much, because it is too strongly orientated by traditional scientific disciplines such as physics, in particular different forms of mechanics. In order to be able to understand technological research as a hybrid of natural science and technology, the concept of the laboratory science is to be confronted with the technology-scientific model of the implicit knowledge, in order to develop a spreading concept in the following scientific research studies.

I. Laboratory science as the constitution of a Research Practice

The concept of an anthropology of science results in the fact that instead of the term of scientific truth reliability and trustworthiness play a central role (Latour/Woolgar 1986, 191). Central points for the description of a laboratory and its possibilities are proficiency, work habits and apparatuses, which are at hand (Latour/Woolgar 1986, 56). Scientific data is not gained by reading nature, but designed during a process of laboratory procedures. Also Knorr-Cetina, following Kuhn's scientific theory, considers the question of theory progress, facticity and objectivism within an historical or socially relative framework. It starts from the point of Pragmatism of knowledge production (Knorr-Cetina 1991, 18f). The regularity of sciences are rather transfactual. They are a product of a process of knowledge construction in the laboratory. Nature and reality are the laboratory construction of Interpretation (Knorr-Cetina 1991, 21f). For an attempt in the research practice, success is more important than a truth criterion.

The production of Knowledge is decisive. Several levels of selectivity can be differentiated. Since knowledge production and laboratory selection show a complex internal structure alternatives and deconstructions are possible. The laboratory situation belongs in very different ways to the context of discovery and justification (Knorr-Cetina 1991, 27f). There takes place the selection of certain interpretations, methods, procedures, in the sense of answers to certain circumstances. This then confronts the question with a context-depending variability, which the sociologist cannot finally determine. Knorr-Cetina considers interpreting and understanding as universal. Also in natural sciences there are background assumptions. Also scientific practice is characterized by interests, interpretations and communications. Terms of everyday-observation are included in science. To that extent, constructional processes would have to be consulted instead of descriptive operations as the basis for the scientific interpretation processes. In addition the opportunistic logic of the production of knowledge is to be considered. The interpretive component of Scientific Rationality would have to be considered more than it is now (Knorr-Cetina 1991, 271f).

The turn to an experimental world view in the natural sciences required a change in the way humans regarded nature. First this had nothing to do with the structure of nature or the peculiar way to explain it, but at first it referred to the boundary line between what was considered as natural and what was not considered to be natural, and therefore considered the question of what is relevant for observation and what is not relevant (Tiles 1993, 468f). The analysis of use of the experiment begins with an examination of the ways, in which instruments in the experimental field of work function. It is difficult to design a taxonomy which adequately reflects the variety and the possibilities of the forms, in which instruments in the experiment appear. This is partially justified by the fact that it appears problematic to isolate instruments in an experimental enterprise. Some of the cases assume that the tactics of the experimenter depend directly on the instrumental capacities and the responsibility of instrumental attesting seems to be a pre-condition for the experimental enterprises. In this sense the experimenter is predetermined in a certain way by the use of his instruments. In other cases however, it shows a constructional and active role, which the experimenter plays when he uses a tool in an experiment and work out its possibilities. The creation of a market for instruments accompanied by standardisation of technologies and technological artifacts. Repeatability and trustworthiness are key terms in the history of experimentation art (Art of Experimentation) and the analysis of experiment (Gooding et al. 1989, 1-3).

The first laboratory experiments isolated certain phenomena and determined their peculiarities. In the second phase laboratory models tried to imitate certain phenomena. Generally, this way of laboratory research spread in the 18th Century (Gooding et al. 1989, 48). There were many attempts to develop measuring instruments with the help of laboratory models in order to reproduce naturally occurring phenomena. Many of the experimental models at that time where dealing with atmospheric electricity (Gooding et al. 1989, 52). The interaction between instruments and scientific research and/or discovery took place in both directions. Progress of instruments led the construction of instruments (Gooding et al. to new discoveries and in reverse affected new discoveries 1989, 58f). Recent work on the scientific theory in principle questions the widespread acceptance of a theory of dominance in the sciences. This is associated with the reproach that traditional science theories considered experiments as test-instances that play the role of verifying or falsifying hypothesizes. The objection says, that thereby neither the self-dynamics of the experimental process of modern sciences, were brought up for discussion nor the relationship of laboratory events to its instruments and in the broader sense to the technological, industrial and social development.

The laboratory is a cultural institution with historical facts. It is based on a much more simple idea than the experiment, and many experimental sciences are not what one could call laboratory sciences. Experimental sciences stand in the context of justification of a certain scientific knowledge. Laboratory sciences produce pictures. Compared to the experimental sciences laboratory sciences are of a larger degree of Technologisation, i.e. the expenditure of technical apparatuses especially in their cross-linking is clearly larger than in experimental sciences. In addition, laboratory science and experimental science have both a slightly different theoretical bias (Pickering 1992, 33-36). One is aimed at the authentication practice of scientific knowledge, laboratory practice deals with production of phenomena, investigation objects, questions and problems up to modified or new organisms. The description of experimental processes has for a long time been operated in such a way, as if experiments have much in common. The way, however, to describe laboratory reports is culturally differently introduced and depends on other defaults, is modified or strengthened in comparison with experimental reports, in a way, which varies from discipline to discipline.

Laboratory sciences can be justified by remarkable success, which evolves when we convert their results into practical action. When prototypes of laboratory science become industrial machines or medical apparatuses they work trustworthily in everyday life, in a world, which has not yet become a laboratory itself. Since this could be coincidence, success or failure in such a mission is insufficient means to verify or falsify a theory which can use phenomena produced in laboratories. With such an assumption the applicability of laboratory sciences would be nothing but a fortuitous event or a wonder. However the application of laboratory science to parts of the world is not completely problematic and not entirely reached by wonders, as one could subordinate, but actually it is a relatively difficult and justified thing (Pickering 1992, 58-61). This is important for the evaluation of genetic engineering exposure tests.

The new methodology of the experiment e.g. the laboratory sciences goes from the paradigm of knowledge to practice. This implies the danger of a Socialisation of the scientific theory. But the central issue is the kind of methodology of scientific culture and practice. Scientific practice must present and interpret itself. The laboratory goes beyond the experimental function and stands for the more comprehensive instrumentalisation and mechanization of empirical research. The laboratory becomes the place of research and the development of prototypes. The scientific technology, medicine and natural science will become laboratory sciences and later we experience an advancement in Technologisation. However, laboratories no longer only generate phenomena, but also material things such as organisms or idealistic things such as methods or test procedures.

The basis of technological practice is developed on the tradition of the laboratory, which changes this paradigm however by the modern means of information technology. To that extent the traditional model of laboratory sciences is fundamentally transformed. However, laboratory practice by itself has no power of validation. It requires a scientific or at least a technological theory. Laboratory practice is an institutionalized form of experimental practice. Justification and validation are no sole formal methodological operations, but institutionalized or quasi institutionalized definitions of the Science Community of methodological standards considering the appropriate questions.

The experimental science is technically oriented; it is instrumentally drawn and therefore has to be accompanied by a hermeneutical-interpretive practice. With laboratory science all these traits are strengthened. This is a grown practice, seen as a whole it is known as the technological variant of the experimental science. In laboratory practice a stronger regress/recourse to the implicit knowledge takes place by cross linking experiments, testing and designing not in order to justify any kind of knowledge but practice justifies itself at least pragmatically with the success of laboratory action. The test character of the experimental conception is being shifted to the success criterion of the laboratory practice, which is shown by its consequences, in addition, a scientification of this implicit knowledge by the continued technological reflection is accomplished.

II. From the Mendelgenetics to Synthetic Biology: Technologisation of a Research Practice

The Mendelgenetics was an genetic experiment but it did not deal with laboratory genetics yet. Before research on Drosophila genetics had almost no access to the laboratory, but instead the garden can be considered as a laboratory. On the basis of a description of the early history of genetics in Great Britain one could consider scientific activity as open-ended and constructive. One could assume that this phase is sufficiently described in the categories of normal sciences. In reality this is a phase of big confusion. Around 1912 two rival theories for the association of hereditary factors were circulated. In order to understand the alternative position of Bateson and his groups both the empirical work of the experimenters and the social structure of his circle and Bateson's theoretical methodological projects must be included (Rheinberger/Hagner 1993, 142-146).

In the thirties a new biology had been developed. In this phase the transition from experimental science to laboratory science in genetics was prepared. However during this time, one can neither speak of a laboratory practice nor of genetic engineering in a closer sense of the word. New biology became known under the name of Molecular Biology. This molecular vision of life was expected to function as a kind of Ockham's Razor. Molecular biology was based on the protein paradigm. Molecular biology defined the level of life phenomena which was set on the submicroscopic range between 10 (-6) and 10 (-7) cm. The new molecular-biological laboratories developed an impressive monumental technological landscape. Research problems were often defined by the technological aids, which were intended to examine these landscapes.

An important step on the way to the deciphering of the double helical structure of DNA was the picture of Roentgen diffraction, which brought a spatial picture of DNA and pointed out the regular structural components. Watson was a biochemist, Francis Crick a physicist, who controlled the Roentgen diffraction. So both scientists combined could recognize the fundamental principles of this structuring. The last crucial insight into the double helical structure was brought by the building of technical models. The two authors made a simple spatial model of the DNA structure, like the masters of cathedral construction made models of their domes, in order to find out, which model could be proven as load bearing. Ten years later it occurred that technological grip (hand grip) was much better than high-complicated computations, which were only 10 years later able to show the regional structure of the DNA. This important role of a technical model foiled the understanding of science at that time and led to the fact that Watson and Crick were not recognized for quite a long time by their colleagues.

With the transition of the modelling of a single gene to the recombination of genes, the transition to synthetic biology is carried out connected with the employment of artificial genes. The model of experimental science is at the same time being transformed to the laboratory practice of genetic engineering. Analogous to the development of synthetic Chemistry by the synthesis of urea by Friedrich Woehler in the year 1828 (Winnacker 1993, 121f), genetic engineering as a new conception and optimization of genes does not understand itself any more as some kind of reproduction of nature. It develops its own artifacts with not only theories and hypothesis, with but with technological means. Thus genetic engineering is no longer a natural science, which discovers laws. However, Ludwig Winnacker is of the opinion that synthetic biology enables a description of nature with more effective methods and opens a new dimension of the mutation analysis (Winnacker 1985b, 19). The optimization of the biosynthesis of antibiotics and protein engineering leads to a pure new synthesis of enzymes on the drawing board (Winnacker 1985a, 20). Simulation and (Re-) construction are characteristics of synthetic biology as a new way of thinking and a new kind of action. Gene technology thereby proves to be a hybrid of natural science and technological research. It is defined as a new type of knowledge and action, which will only unfold itself fully in 21th Century, as the new research practice.

The modern character of genetic engineering is not to be found in a new type of explanation. It lies rather in a specific development of experimentation art, which has been described under the key word of Technologisation. Thus, the type of knowledge in the sense of a change from knowledge of something <<Wissens von etwas>> to the knowledge of dealing with something <<Wissen wie damit umzugehen ist>>, not however the paradigm of experimental research. Thus, experimenting becomes the new form of knowledge (Zimmerli/Hohlfield 1991). The completion of the modern science paradigm is at the same time its transformation. This articulates itself in a technologisation of empirical sciences where computer simulation tends to take over the role of the experiment. Following Walther Zimmerli technologisation is to be understood in that way that scientific basic research, technical application and economic use can no longer be separated sharply, but merge into a type of scientific action. Secondly science becomes more and more technological and technology becomes more and more scientific (Zimmerli 1993, 299).

Since the middle of 70s a new research practice is being constituted. What is left in the traditional idea of experimental science with its conception of scientific laws is the research practice. They are replaced by technical rules. The difference between nature and technology becomes blurred ever more. Research practice produces a method of dealing with knowledge in the laboratory, in field research, in agricultural sciences and in biotechnological breeding practice. Thus an extension of traditional laboratory practice takes place: Due to the risk problem, and during the release of transgenetical organisms, step by step, an expansion of laboratory research is carried out into field research. The entire earth becomes a potential laboratory. Constitutional is a research practice, not space – even if this is an S-4-Lab in Germany. So far the dealing with knowledge acquired in a laboratory with its regularities proved to be compatible with the dealing with knowledge in nature. Analogy conclusions are allowed, because they are based on pragmatics. Fundamental differences between the artificial laboratory and the naturally oriented laboratory are hypothetical and artificial. Therefore, pointing out the constitutional elements of the pragmatics of genetic engineering appears to have priority in order to be able to reveal the leading idea.

The fact that life sciences and general science are dependent on technology does not seem to be controversial. However the importance of this fact from a philosophical or historical point of view is an open question. For most philosophers and historians of science the necessity for a consideration of technology is epistemologically irrelevant. Technology is regarded as a component of the discovery coherence of our knowledge about living nature, without any consideration for the coherence of justification of this knowledge. However, far more important than this controversy is, from the point of view of Philosophy and History of Science, the fact that some of the main structures of science are structures of action, which are determined by the acquired structure of research technology.

The subject of genetic engineering is primarily the molecular basis of genetic information for the development of certain physical characteristics. The developing body of an organism can be described with the terms shape and structure, the development is a process of cross-linking, how it has to describe the molecular biotechnology. Technology is a mixture of actual technological Know-How and technically instrumental products, mixed with methodological knowledge to a procedural and experimental-based Know-How, whose actual point does not represent a bare factual knowledge. Moreover it includes an active access to certain components by methodical work. It is a question of acting with a certain understanding. And this is an acting knowledge; an implicit knowledge of regularities of the development of organisms, which gradually develops through technical handling of genetic information.

The approaching point of genetic understanding and genetic reconstruction are the smallest units of life in their structures. This means, the object of biological understanding is not only the total organism in its execution, but has to take place on different levels in the sense of networks and their structures. To that extent, a critique of Holism is also justified, not only a criticism of reductionism. Thereby characteristics can primarily have physiological functions, in addition to physical and behavioral characteristics. This means that in this case different constitutions of objects of biological science are necessary. It concerns the understanding of information, which is contained in molecule bonding, in order to be able to deal meaningfully with it. In this case there is no reductionism and also no `layer ontology' but on the contrary different situative contexts for biological constitution are being worked out. The Structures must obviously be examined in different ranges or levels of network. As it has been pointed out at the end of the first chapter, (1) the Genetic Code, (2) the Level of the Proteome, (3) Cells, (4) Organs, (5) Organisms and the (6) Ecological systems for a Hermeneutics of biology could be differentiated. Different types of object of biological understanding, i.e. also different levels of the constitution of objects in the subject of biological science, can thus be differentiated.

Because practical research work in the laboratory or in the field represents a kind of handicraft, experience knowledge, the 'knowing how', is also an important factor for the success of such work. Uncertainty of tasks is hereby characteristic (Hasse et al. 1994, 180). With the context of discoveries leading conceptions are the central issue. Also the allocation of scholarships determine the practice by the choice of topic in a crucial way, in particular by the structure of three-year projects. The financing of a final thesis with the help of a third partner for the project lets students and Ph.D. students participate in the research, whereby due to the temporal limitation topics with a minimum of uncertainty are being placed (Hasse et al. 1994, 184). Research objects are basically transgene useful plants. particularly Important is the choice of an organism model and a model system for research. Also larger research programs have a model-character (Hasse et al. 1994, 197). A whole branch of industry is concerned with the production and supply of instruments, devices and chemicals for cell research and for molecular-biological research, which thereby contributes substantially to the standardisation of research in this area. Within this range also the costs of mechanization manifests themselves. In the area of molecular biology practical and experimental-based work becomes generally accepted, however thinking work is not redundant. However, there is an increasing demand of interventionist knowledge in this research. Research activity has a preliminary structure not least by the range of recognized methods. Important is the strategy of „Educated guess" (Hasse et al. 1994, 207). (This „Educated-guess") This leads to an extension of the classical genetic methods by methods of Molecular Biology (Hasse et al. 1994, 212).

The Technologisation of biotechnology is based on the development of procedures, in particular on test procedures for the use in professional as well as in non-professional everyday life. Test procedures are however no more experimental arrangements, but technical guidances. In scientific theory, biotechnology is not only a new sociology of science, but it is a new methodology after the pragmatic turn in scientific theory, which in turn failed more comprehensively than originally intended. Despite all pragmatism and sociologisation a methodologically oriented turn remains. In the Technologisation of biotechnology it comes to an expansion of the instrumental dimension and to the constitution of research defined as a practice. Technological artifacts are no longer being produced – but services, test procedure, model organisms, Know-how, genetically changed organisms up to synthetic life are being produced. Test procedures and reconstruction of organisms are no experimental activities, even if they are being followed in a certain way according to the scheme of Trial and Error Method. The success of a technological construction depends on completely different rules than the success of an experimental test procedure. Experiments want to grasp the given; in this aspect they are still similar to test procedures. Construction aims at new combination of instrumental grasping of objects. It is not the process of an experiment which is being constituted but the object of research itself, the organism.

Genetic Engineering as a laboratory science covers the experiment, the isolation of its object of study (the gene), the new combination of the isolated one and the reconstruction of organisms or parts of organisms. Laboratory tests are no longer only existing, rather in the case of the genetic engineering it undertakes the step into synthetics - both in biology and regarding the organisms.

III. Methods and Technical Procedures of the Genetic Engineering: Technologisation of a Laboratory Practice

Following a definition by Ernst Ludwig Winnacker genetic engineering can in general be understood as procedures for the isolation of genetic material, in order to build new combinations of genetic material and to introduce and reproduce re-combined nucleic acids in a possible new environment (Winnacker 1985, 14). Primarily, it is interpreted as a method of basic research, which illuminates the structures of genetic information and its expression. Their methods serve the development of fundamental knowledge regarding the connections between structure and function of proteins in different organisms, and for studying the expression of Genes and the development of organisms. If one understands technology as measures and procedures, which are consulted by human beings under the utilization of laws of nature and materials, in order to make it usable for production, then genetic engineering is the in-vitro-production and in-vivo-reproduction of recombinant DNA (DeoxyriboseNucleic Acid). Technology as methodical teachings and procedure science is the teaching of the totality of procedures, processes, methods and the production procedures of certain products (recombinant DNA in new cells). If one proceeds from this definition it would have to be considered as genetic work procedures in the context of biotechnological methods for the multiplication of products and knowledge. Genetic engineering could then be considered as that area of biotechnology, which used genetic procedures at least in one of its steps.

In the years 1971 to 1973 a new laboratory technique in the form of DNA recombination technique and DNA cloning was developed. Vectors of the Cloning of DNA belong to the new tools of genetic engineering. Tools of genetic engineering usually are technically changed bio molecules, in exceptional cases also the classical and technical instruments such as Pipettes or electrical impulse in Electroporation as well as in different techniques of the microscopic view. The methods and the group of techniques where being combined as a kind of technology in the sense of a systematic procedure. Investigating the position of the cloned gene leads to the localization of a cloned gene in a large DNA molecule. For the determination of the position, a hybridization process is necessary. The process of hybridization (the Blotting procedures) serves the clearing-up of structure, organization, function and expression of genes. These serve a recontextualization of genes in the context of a molecular object-constitution of genes.

Such genetic and biotechnological methods are: (1) production of recombinant DNA molecules. This is done via cutting of DNA molecules in defined places by the restrict endonucleus process and by connecting different DNA molecules by means of Ligasen. (2) Insertion and Cloning of DNA molecules into cells. Insertion is done by transformation and by means of so-called vectors (viruses, Plasmids) or by micro injection. In addition there is the establishment of DNA in the Genome by the integration of vectorial DNA in chromosomes and by homologous recombination. (3) Sequencing of DNA and proteins in order to gain information on principles of the Gene-expression and gene-regularisation and on DNA- as well as in protein structures. (4) Amplification of specific DNA sections with little pre-information by polymerase-chain-reaction (PCR), a biotechnological procedure, which is based on genetic engineering. (5) different techniques for the identification of DNA and proteins with marked nucleic acids and/or anti-bodies (so-called Blotting procedures). (6) Chromosome-Walking in order to find unknown genes. (7) artificial synthesis of DNA, nowadays largely done by machines. The methods (3) to (7) are not genetic procedures in the actual sense, but biotechnological methods, which developed out of genetic engineering. At this junction, a sharp demarcation (delimitation) between genetic and biotechnological procedures is not possible.

The new laboratory technique can be designated as the revolution of modern biology. A completely new methodology is developed, with which one could plan and implement experiments, which had previously been impossible in the laboratory. Experimenting was not free of problem, but at least it was successful. The principal item of these methods, which in summary can be called DNA recombination technology or genetic engineering, is the procedure of DNA cloning. This introduced the third major age of genetic engineering. Until today we are experiencing the impetus released by this revolution, and so far there does not seem to be an end to the exciting developments. A central component of each genetic transfer is the vector, which the gene transported into the original cell and there ensures its multiplication (Brown 1997, 15f). Although DNA Cloning is still a relatively new procedure, it has already developed to a most sophisticated technology. Today a broad spectrum of different Cloning-vectors is available. They all come from natural plasmids or viruses, which however were usually changed in different ways. Therefore each of them is suitable to clone a very certain type or gene. The cloning of a gene is a relatively simple procedure. Nevertheless it is of importance, because one can gain an individual gene in pure form by the process of cloning. It is separated from all other genes, with which it is normally presented in a cell (Brown 1997, 18f). Plasmids were the first starting points for the construction of vectors (Brown 1997, 23).

Molecular biology is not a traditional natural science. So far it basically has made progress by the mixture-of-methods (Methodenmix) and by new technologies. The result is qualitative information about the materialistic information, but not quantifiable knowledge of law. Important for molecular biology was progress in the understanding of methods. However, biology as a comprehensive understanding of structure and function of the entire cell, has not yet been replaced by molecular biology by any means. Molecular biology is a milestone on the Way to Technologisation of science offering a new kind of explanation. Quantitative aspects are so far neglected in genetic engineering. Thus molecular biology seems to be on the way to becoming a comprehensive qualitative science. In principal genetic engineering has reached the status of a research practice since the experiments done by Khoranas. The introduction of the computer into the modelling of the connection between genomics and proteomics, the whole of the proteins, the further use of experimental procedures and the development of separation and analysis methods speak for the development of a comprehensive laboratory practice, which registers a constantly increasing degree of technologisation.

Many of the outstanding insights into molecular biology during the last 40 years where reached by the illumination of molecular structures. Perhaps, the most spectacular and best well-known insight was the modelling of the DNA Helix by Watson and Crick. In this period, experimental studies were done on over 500 protein structures by the diffraction of Roentgen X-rays in crystals. Computers are now the mandatory instrumentation for the determination of molecular structures and are not only used in the collection and computation of data. They are also used to produce folders for the density of electrons, to illustrate these folders and to implement amino acids into these maps of electronic density. Moreover computers are used for the final process of compilation. Besides that there was a massive increase in computer techniques, which are used to point out macromolecular structures in its connections, its dynamics and its reciprocal effect.

IV. Synthetic Life - Perspectives towards Genetic Engineering as a Science of Construction of Organisms

With the first forms of completely synthetic life a limit will be exceeded, which, however is only a limit regarding the past technical know-how. The contrast between nature and technical skills is proven to be wrong, equating of nature and the morally good as well as technology and moral outrage has already been implausible. The discussion of artificial human beings is re-animated and might be focused on really artificial humans and completely new forms of the man-machine interaction. For the theory of dealing with technical action, there is no given limit of technical action outside the laws of nature, but different forms of the not-yet-know-how. Everything that is testable has to be tested, as long as it is pragmatically and ethically justified. Then it requires the mathematical simulation and modelling as well as a hermeneutical interpretation. Ethical reflection is different from technological testing but not independent from it. Synthetic life in its more simple forms will probably be ethically justified. The possibility of being able to create new synthetic forms of life in the foreseeable future might seem to be a serious trespass beyond the current limits. However, this process reveals only the dynamism of technical action between the technological imperative and ethically reflected action.

The technological construction of artificial forms of life requires a considerable degree of scientific and technical know-how. However it has to be assumed that at least in the beginning artificial forms of life will not fully lose the character of black boxes. One will have to try things out when sketching artificial life and also commit risks. But a high measure of knowledge will be necessary for technical designing of organisms, probably more than with genetic manipulation of an organism. The practical value of technically designed organisms results from the form of life and organization of an organism designed according to its use. However, not all possible forms of use could be determined in advance for all forms of synthetic life. An organism is designed as an instrument and is not being instrumentalised afterwards. However, the question remains whether this, as different forms and degrees of instrumentalizations, will affect the ethical evaluation. For a re-synthesized organism there is no life appropriate to a species, which would be independent of humans. Criteria for the conditions of keeping this organism could only be „derived" from its organization form. First however it must be expressed explicitly, which conditions must be fulfilled for the technical construction of artificial forms of life, in order to be able to speak of a completely artificial manufactured cell. A comparison has to be made between the reconstruction of an artificial organism and a technological reconstruction of a technological means (invention). This requires an analysis of genetic construction of action. A condition for this is the progress in the postgenomical era of genetic engineering. Functions of genes and intra-cellular networks must be known better.

Synthetic life is based on a complete synthesis of also the given genotype of organisms, which has not existed before. These are completely synthesized micro organisms or radical reconstructions of mammals in the sense of the chimaeras or hybrid formation, i.e. the process of cloning, which do not permit any longer a clear kind and generic affiliation statement. New forms of synthetic biology or molecular bio-engineering require a comprehensive knowledge of potential abilities of a cell inclusive that of proteome and transkriptome analysis as well as the comprehensive modellings of bio computer-science. New minimum organisms can be designed regarding a potential specific use. A specific organism can be designed regarding its function and its potential of utilization. As organism it will also claim patent worthiness (because it is now artificial) and it can be used in basic research, in the production of new materials and in medicine or pharmacy. A completely new way of bio-engineering is about to be developed.

References

Brown, T. A. 1997: Gentechnologie für Einsteiger. Grundlagen, Methoden, Anwendungen; Heidelberg 1997

Gooding, D. et al. 1989: The uses of experiment; Studies in the natural sciences; Cambridge

Hasse, R. et al. 1994: Die Technologisierung der Biotechnologie am Beispiel der Pflanzengenetik und Pflanzenzüchtungsforschung; Abschlussbericht zum BMFT-Projekt „Die Technologisierung der Biotechnologie: Zur Durchsetzung eines neuen Wissenstyps in der Forschung; Mass. Erlangen Nürnberg

Hausmann, R. 1995: ... und wollten versuchen, das Leben zu verstehen ...; Betrachtungen zur Geschichte der Molekularbiologie; Darmstadt

Irrgang, B. 1997: Forschungsethik Gentechnik und neue Biotechnologie. Grundlegung unter besonderer Berücksichtigung von gentechnologischen Projekten an Pflanzen, Tieren und Mikroorganismen; Stuttgart

Irrgang, B. 2001a: Technische Kultur. Instrumentelles Verstehen und technisches Handeln; (Philosophie der Technik Bd. 1) Paderborn

Irrgang, B. 2001b: Gefangen in Sachzwängen? Zur ethischen Dimension der Gestaltbarkeit der Biotechnologie; in: St. Heiden et al. (Hg.): Biotechnologie als interdisziplinäre Herausforderung; Heidelberg/Berlin 2001, 83-96

Irrgang, B. 2002a: Technische Praxis. Gestaltungsperspektiven technischer Entwicklung; (Philosophie der Technik Bd. 2) Paderborn

Irrgang, B. 2002b: Technischer Fortschritt. Legitimitätsprobleme innovativer Technik; (Philosophie der Technik Bd. 3); Paderborn

Irrgang, B. 2003: Von der Mendelgenetik zur synthetischen Biologie. Epistemologie der Laboratoriumspraxis Biotechnologie; (Technikhermeneutik Bd. 3); Dresden

Irrgang, B. et al. 2000: Gentechnik in der Pflanzenzucht. Eine interdisziplinäre Studie; Dettelbach

Knorr-Cetina, K. 1991: Die Fabrikation von Erkenntnis. Zur Anthropologie der Naturwissenschaft; Frankfurt (1981)

Latour, B., St. Woolgar 1986: Laboratory Life. Reconstruction of Scientific Facts; Princeton N. J. 1979

Maddox, J. 1992: Is molecular biology yet a science? ; in: Nature 355, 201

Pickering, A. 1992: (Hg.) Science as practice and culture; Chicago, London

Rheinberger, H.-J., M. Hagner 1993: Die Experimentalisierung des Lebens. Experimentalsysteme in den biologischen Wissenschaften 1850 bis 1950; Berlin

Tetens, H. 1987: Experimentelle Erfahrung. Eine wissenschaftstheoretische Studie über die Rolle des Experimentes und der Begriffs- und Theoriebildung der Physik; Hamburg

Tiles, J. E. 1993: Experiment as Intervention; in: The British Journal for the philosophy of science 44 (1993), 463-475

Winnacker, E. L. 1985: Grundlagen und Methoden der Gentechnologie; in: Max-Planck-Gesellschaft München (Hg.): Gentechnologie und Verantwortung; München, 14-21

Winnacker, E. L. 1993: Am Faden des Lebens. Warum wir die Gentechnik brauchen; München

Zimmerli, W., R. Hohlfeld 1991: Interdisziplinäre Technikfolgenforschung; IGW-Report 5, 79-85

Zimmerli, Walther Ch. 1993: Die Bedeutung der empirischen Wissenschaften und der Technologie für die Ethik; in: A. Hertz, W. Korff, T. Rendtorff, H. Ringeling (Hg.) Handbuch der christlichen Ethik; Freiburg, Basel, Wien, Aktualisierte Neuausgabe Bd. I, 297-316

Justified Trust in Technology

Bernhard Irrgang

A hermeneutical concept of technological knowledge and understanding proceeds from the implicit knowledge and develops on this basis, an understanding of technological action, on a phenomenology of the tool use i.e. dealing natural processes, machines and technical systems in instrumental understanding as an implicit handling knowledge or competence (Irrgang, 2001a). Not the analysis of tool is of priority of importance, but those of the success, which can be achieved by means of a technical means, within the context of technological practice. The kind of realization of an intended effect is investigated. The interpretation of implicit technical knowledge is based on Martin Heidegger's existential analyses of technological handling with the ‚Thing-world' in „Being and Time (Sein und Zeit)" (Corona/Irrgang, 1999). The concept of an implicit handling knowledge due to an understanding process of ranges and application of natural processes or of tools becomes starting point for a philosophization over technology. This implicit handling knowledge should be reconstructed in the sense of a nested interweave knowledge and doing. Certainly by the special structure, with which one goes around, and habituality of the going around. Only in second line are decisive for technology explicit knowledge, mathematization and scientification (Irrgang, 2001; Irrgang 2004).

As tools and machines became autonomous, the more represent them in an incorporated technological knowledge, is later called as research knowledge. Thus, the technical authority and competence of technical means developed. But, they are never only autonomous, rather ever with respect to the horizon of a technological practice. It is still in more or less strong dependence on the technical authority and competence of technician in the course of history of technology, apart from technical authority of technicians. Of it is differentiated as a technological reflection knowledge (Meta Technical Science). The meaning of technical artifacts in their cultural imbedding concerns and extends over the functional certainty of a technical artifact around user knowledge and dealing with knowledge.

I. Trust in technology through the scientification of construction?

Since the processes of industrialization in 19th century, it justifies itself the trust in technology no more on the authority and competence of craftsman, but by scientfication (Verwissenschaftlichung) or professionalization. First by any means, the proof did not help superior efficiency in the construction practice and scientific methods to the break-through. Their meaning was rather in those few cases, in which up to build and machine-mechanical analyses had found into the construction process entrance, rather small. The construction practice still got along at this time without scientific procedures, which anyway only ex post were used to the science-supported analysis. Besides be in view of both still small relevance build and mechanical procedure including their mathematical apparatus and insufficient develop knowledge the participant with the new theory engrave error with mechanical beginning and in method of calculation to the agenda. To that extent now the example drawing into the everyday life of designing attained scientific engineer activity the action-oriented strength and shapes in the association with the penetration of school outline, in the <<maschinenwesen>> and in the technical occupations of building industry rapidly, the professional identity of engineers, was able against it by no means action-determining to work. For the society science turned out primarily to an instance to produce those in technology convertible safe knowledge has (Haenseroth, 2003, P. 21).

The paradigm of alleged infallibility of scientifically justified statements ensured for the fact, that into the broken open gap of the apology and authentication of new technology now apparently objective, generally accepted scientific knowledge grew as abstract truth proof, of a „objectively correct" supported of authority of the science, as it were specific obligation-like technology. From the confidence and trust into the progress of science as foundation were able themselves to present modern progress faith now nature and technology-scientific results and realization systems for compensation all these deficits of authentication and justification (Haenseroth, 2003, pp. 22-24). Engineer-moderate self description and foreign description of engineers fall apart. Because epistemical security may not be confounded with technical security.

II. Maintenance and technical security: institutional reason and foundation for trust in technology?

In the modern technology, the life span of a construction is usually given. Modern constructions are therefore constructions based on time. The term of life span of a construction unit is ever more frequently defined than the assignment, which a construction unit up to technical bears an incipient crack (Broichhausen, 1985). Causes of deviating the technical artifact from specified condition are the cause for maintenance, whereby as partial measure of the maintenance expectation, inspection and maintenance can be differentiated. Simon claims a categorical unit of production and maintenance of technical artifacts. The technology-immanent and a technology-transient maintenance are to be differentiated. The fitness duration, the service life and preservation duration are likewise different aspects of maintenance. The lacking of maintenance shows up frequently with the study of causes of technical disasters. To that extent, there is a responsibility for maintenance and institutions of the maintenance. Therein manifest, in particular the economic aspect of maintenance as well as the dimension of technical security during the maintenance. With the action type of retaining is to be likewise proceeded from the irreversibility of acting. Simon spells of an opposing of target (Simon, 2002).

Heidegger speaks of the <<Hinschwinden>> of reliability of technical artifacts, wear and consumption mentioned. Extreme events can be likewise understood as causes for the destruction and the fall technical artifacts. Causes for maintenance efforts are deviations of the technical artifact from the specified condition. Natural, social, economic and technical disasters can be differentiated. In the purge the technical artifact is over-formed by nature organization or destruction. The repair and maintenance cause a subsequent stronger individualizing of these <<Plurikate>>. Maintenance obligations are expression of the historicity of technical artifact. It manifests itself in fitness duration and service life. At the end, a value decision stands the renouncement of further maintenance. Altogether a total expenditure can be intended for the use of an artifact (Simon, 2002).

III. Limits of the controllability with modern technology

In 20th century the power of our machines and our inventions dramatically increases, at the same time in addition, their foreseen consequences also remain intact. Modern technology is fundamentally different from earlier forms of technology. The principal reason for it, is their complexity. Complexity creates uncertainties, limited what we can know or reasonably consider over a technology and its development into the future (Pool, 1997). One hundred years ago, Americans considered principle technology as a good thing. Today there are countries, in which caution plays a central role and basic attitude in relation to technology. In some countries, this caution and suspecting of technology went so far that certain technical developments were terminated e.g. genetic engineering and core technology. The process of the invention is led of the faith & trust and of practices. These are developed with years of experiences with trial and error before and are resistant in relation to radically new ideas, which technologically open new ways completely (Pool, 1997). Not growth absolutely is technically risky, but a size growth of large technical systems quasi without borders. We need another architectural structure in times of the globalization for the growth of large technical systems as cancer-like rampant growths and coincidental dyes. The more infrastructures we have, the more must we on their receiving end or we give up. In addition, maintenance and repair create jobs.

Out of control to turn out to be able, is a characteristic of complex technology. The fact that something is out of control interests us only, if we expect that there would have to actually be control regarding this development in the first place. Not all cultures e.g. divide our insist on the ability to control things. The modern science apparent brought an increasing of controllability. In conventional perspective works are more than safe of the technology. What humans made can it also control. This is common-sense opinion: Control is altogether seen to part of the construction idea of the technical creation. Tools hang completely off of the will of the using (Winner, 1992).

The controllability thesis of technology is the basis wishful thinking. The possibilities of heavy collapses in high-tech systems are not to be underestimated. The problem of using and controlling places itself also on this level. Which we find with high-tech structures, not the passivity of a tool, which waits for it, is to be used, but a technical ensemble, which requires routine and trained behaviour and acting (Winner, 1992), which must be also acquired in handling. The technical medium changed, it is no more but tools with a more and more own operational character, not however the fundamental structure of technical acting. We are not today in the everyday life any more able to repair all the technical devices us surrounded, but the successful use of a technical artifact or also successful handling technical structures does not presuppose that we would have to know details of the construction plans of technical artifacts and technical structures. Not only an engineer is successfully technically an action. This implies a certain democratization of technological action and the technical development (Winner, 1992).

IV. Technology at the forefront

Human beings, who work on the technological front, require to its self-preservation a higher degree at security. Human mistakes (Irrtümer) and technical malfunctioning is inter-linked. This is not failing at itself, because many safety systems built in redundant way. It is however most difficult and requires sophisticated experience to interrupt the chain from events to which are straight thereby, a potential malfunctioning to release (Chiles, 2001). Both in the aircraft construction and in space travel front engineers in many cases had become victims of their own promises, with which it applied to develop or develop it for the technology to support. In many such ranges seems all too large confidence and trust in technology in tragic form to have been measure-led. One can call this a fatal trust in the technology. Therefore, for the evaluation of a new technology a safety philosophy and safety test of a technical plant are important. We need technology with inherently safe construction (Chiles, 2001).

The rarity of structural errors in the technological construction led to the acceptance that engineering and technological construction do not imply an extra large risk even in a very risky enterprises (Petrowski, 1992). If today still some structures of technical kind exhibit and accidents which lead errors, then that is essentially, because still technological limits are overcome, also in the new industrialized world. 50 to 90 percents of all structural technical errors as a consequence of the attempt is rated to overcome the factor size. A further important factor is the incorrect material. The paradox of technological construction is that successful structural technical concepts in errors can lead, while colossal errors are very important finally, in order to advance the development of innovative techniques and inspiring technical structures. Errors are inherent in all useful technological constructions, due to conflicts between the requirements of the user, which are frequently unknown, and who construction for a use, which is frequently also simply arbitrarily accepted (Petrowski, 1992).

V. The „revenge" of technology (Technological Revenge)

With the introduction of electronic office paper, consumption increased infinitely. One can call these paradoxes as consequences a „striking back of the things" or the „revenge of the things". Exactly taken therein the risks of technological extrapolations manifest themselves. A hope for correcting and repairing the technical world is an illusion. One must turn into from the universal model of the „revenge" to individual feedback effects. Starting point are paradoxes, behaviours of technical objects. Mostly not intended consequences are unpleasant, while the pleasant are most surprising. In addition, we discover also the pleasant positive effects only after negative experiences. New technologies are valuable only if we can put a new habits of use with them. A setback effect is not the same like an individual act of revenge. If for example, the cancer treatment produces a further deadly cancer, then in the strict sense, a revenge effect is present. Security is another idea for revenge effects. Alarm systems lead revenge effects to many wrong alarms and arise, if we institutionalize technologies (Tenner, 1996).

Revenge effects („things bite back") happen or occur, because new structures, tasks and organizations with real people must react in material situations in kind and ways, which cannot be foreseen. Also nature „strikes back". Tool systems can strike back in a certain way. Parts of machines, which do not interact with one another in desired and not foreseeable kind, can lead to such revenge effects. The consideration of the use, the management is important. System effects and handling and dealing effects can interact. In 19th century it gave a technical Prometheus Titanism and a summit of the technological optimism. Actually however it had to be learned at all times from accidents. A too high trust in the technology and in the own abilities leads inevitably to errors. Totally safe and secure technology is already not possible because of technical users. The modern concept of side effects originates in the 19th century. Particularly there are many infections in the hospital? (Tenner, 1996).

Many ranges of technological development have limits of the intensification. Ambivalence and the potential disaster are the character of technology, and need to be emphasized. The fall of Titanic led to the introduction of a distress emergency service. The automobile and the consequences for the safety engineering are to be likewise considered. In all other respects also horses and cars were dangerous. The reduction of setback effects requires high technical intelligence. Technological optimism means the ability in practice to recognize promptly bad surprises. It can be promptly recognized earlier so that, one can do something against it (Tenner, 1996). Edward Tenner describes a fundamental pattern in the logic of technical failure. There are still other ways. This pattern however is as much of special interest as construction, production and need in one another seize here, if the action situation is particularly unclear and so errors can develop easily.

VI. Risk and technical acceptance

Safety and risk analyses must confirm the sufficiently safe enterprise of a plant. And this must be provable not only opposite the responsible experts, but the substantial trains of thought should be comprehensible also for non-specialists. As good a transparency of the system, i.e. trustworthiness of the citizens with respective technology, comprehensibility of the damage mechanisms and controllability of technology as possible are important. Terms of hostility to techniques, technical criticism and technical acceptance are much too general and must be analyzed critically. These are to a large extent inadequate for a discussion of technology evaluation and the technical experience. As alternative terms those are addressed and technological skepticism, whereby the intellectual and the practical levels of technology evaluation must be differentiated again and again. Technology criticism is in the reason society criticism; the attitudes to technology depend on the concrete life situation of concerning. We could constitute different forms of satisfaction of the process movements: Forms of the penetration of new techniques with assistance of law and government authority, over market mechanisms or by incentives (Stadler/Kuisle, 1999).

If we refrain from the term of restricting acceptance, then left themselves are new problem areas (resistance against environmental impact, reaction to industrial accidents and occupational illnesses etc). One pointed out repeatedly that with the problem of acceptance, the action clearance of the concerning is to be considered. In addition, it was pointed out that with the problem of acceptance under circumstances the action clearance of the concerning is to be considered. Here there are overlaps with the analysis of the term of <<zumutung>>. With the term of guess <<zumutung>> are good in particular technology penetration and the resistance situation to characterize. The discussion of the new one, progressive ones of technological innovations, shaped by generally spread growth and progress paradigm, adjusted for a long time enough for us the view to ask also for the quality and the continuance of a function. In particular it must be referred to the time-bound (temporally) character of technology evaluation and the criteria consulted for it (Stadler/Kuisle, 1999).

VII. Innovation and routine: Technical trust due to experience

Technological and scientific innovation accompanies inevitably with uncertainty. Hereby it comes to a false estimations (Rescher, 1983). In view of the common ideology of technological change and technological progress should be also considered, which suffer from innovations. Innovations stretched a field of utopian way of liberty and the ideal planning. This should take place via the market and technological laws, to it lead, which creative philosophers and the freely floating of capital unite, in order to create a new world, the abundance have at technical possibilities and energy. The inconceivable utopia of a technology without innovations was neglected in this contrast. Technology is linked in its internal nature with change (Williams, 2002).

The identity crisis of an engineer, which expresses itself therein that this has no more goals, has to an ideology of the engineer nature led, which identifies the engineer work with scientific methods. The center 20th century was, that is, the period of rapid professional growth of engineers, after that II World war. A more imaginative education of engineers and a new kind of the broader educating movement would have to inter-linked. In addition, a reconstruction belongs to a new technical idea of products of engineer. The technological determinism should break cultural open of resistance. There is the risk of a over enlightenment. Thus, it came to a confrontation of the technological change with the historical change. Technological progress should orient itself at the community. It manifests itself however, in a crisis of environment and in particular in a tremendous scarceness at time (Williams, 2002).

The risk society is however also a trustworthy society: Trust is the dark side of the demystification world - and trusted friend is the side of a natural one, inevitable ground of technical environment. The material history of technology in the modern trend is an amazing triumph– social-critical semantics insists on interpreting these successes as marginalization of non-technical communication instead of asking for the trustworthiness of technical discharges. How is it thus possible that in everyday dealing with technology completely nonspecific trust achievements are applied with striking without efforts? It meant anyhow technical artifacts, thus technical articles, are even things, which function in contrast to techniques in the sense courses of action easy to learn and led by rules. Thus, typically usual consumer technology, technology for laymen goes around technical things in the form of already imported, practically experienced things, how it is daily available. It concerns the trustworthiness of technical things. It should be asked rather why despite the often experienced potentials of dangers of technology and its social costs, our society–comparably, that, on which Weber called as „Agreement on Community" aimed–as trust community in technology can be designated. Approaches of a standard format of the technology in its functioning are to be developed further. But, in the linguistic use of Heidegger–uncanny one of the technical world is however, nevertheless functioning of technology and not their unexpected failure (Wagner, 1992).

The approach towards technical trust as authority of confidence/trust or the modern society as technical trust community can be found in Max Weber. Weber differentiates agreement into a rational order (it is technical, social or symbolic kind) of the understanding of its complexity: Agreement and understanding are not identical. Will agreement-act of Weber as habitual action is classified (Wagner, 1992). The more we surrounded by technical artifacts, the smaller becomes the cut-out of an understanding-moderately comprehensible and the more largely become the functionally necessary need of trust. Each attempt to come to actual special control throws new hidden contingencies and latent uncertainties (Wagner, 1992). Trust in technology is produced not by ethical consideration, but by the everyday use. Therefore, expert techniques such as nuclear technologies or gene technologies do not easily have it.

Trust in technology is a thing to deal with knowledge, as tacit knowledge. One cannot produce it by proofs and authentication strategies before the use. One can persuade at best to try dealing out nevertheless once by use. It is finally a question of routine, the structure and learning routines, proven handling technical artifacts, which create trust in technology. One cannot force trust in technology under any circumstances. The repetition of results, the routine and the probation lead finally to confidence in technology. The thought of control of technology, in the sense of the control of a thing, is not correct in principle. Rather controlling own authority goes around and learning. The reproduction of a successful conclusion of an experiment in the sense of the probation creates trust. It is thus the repetition, not the rational argumentation, which creates trust in technology. If one gives at all no possibility to a technology to be able to work satisfactorily then one never gave the chance to this technology to be able to prove their legitimacy. Such pre-prohibition is dogmatic, when if so, it speaks good reasons for an extreme danger of a certain technology.

References

Broichhausen, Josef 1985: Schadenskunde. Analyse und Vermeidung von Schäden in Konstruktion, Fertigung und Betrieb, München/Wien.

Chiles, James 2001: Inviting disaster. Lessons from the age of technology; New York.

Corona, N.; B. Irrgang 1999: Technik als Geschick? Geschichtsphilosophie der Technik [Technology as Destiny?: History of Philosophy of Technology]; Dettelbach.

Hänseroth, Th. 2003: Die Konstruktion „verwissenschaftlicher" Praxis: Zum Aufstiegs eines Paradigmas in den Technikwissenschaften des 19. Jh. [The Construction of „Scientification" of Praxis: To the Development to a Paradigms in the Technological Sciences]; in: ders. (Hg.) Wissenschaft und Technik. Studien zur Geschichte der TU Dresden, Köln u. a. 15-36.

Heidegger, Martin 1972: Sein und Zeit; 1972; Tübingen.

Irrgang, B. 1996: Von der Technologiefolgenabschätzung zur Technologiegestaltung. Plädoyer für eine Technikhermeneutik; in: Jahrbuch für Christliche Sozialwissenschaften 37, 51-66.

Irrgang, B. 2000: Technological Development and social progress; in: Instituto del Filosofia Pontificia Universidad Catolica de Chile; Seminarios de Filosofia 12/13 (1999/2000), 41-52.

Irrgang, B. 2001: Technische Kultur. Instrumentelles Verstehen und technisches Handeln [Technological Culture: Instrumental Understanding and Technological Action]; (Philosophie der Technik Bd. 1) Paderborn.

Irrgang, B. 2002a: Technische Praxis. Gestaltungsperspektiven technischer Entwicklung [Technological Practice: Design Perspectives and Technological Development]; (Philosophie der Technik Bd. 2); Paderborn.

Irrgang, B. 2002b: Technischer Fortschritt. Legitimitätsprobleme innovativer Technik [Technological Progress Legitimacy Problem of Innovative Technology]; (Philosophie der Technik Bd. 3); Paderborn.

Irrgang, B. 2004: Konzepte des impliziten Wissens und die Technikwissenschaften [Concept of Implicit Knowledge and the Technological Sciences]; in: G. Banse, G. Ropohl (Hg.):Wissenskonzepte für die Ingenieurpraxis. Technikwissenschaften zwischen Erkennen und Gestalten [Concept of Knowledge for the Engineers Praxis: Technological Science between Knowledge and Design]; VDI-Report 35; Düsseldorf 2004, 99-112.

Irrgang, B. 2006: Technologietransfer transkulturell. Komparative Hermeneutik von Technik in Europa, Indien und China [Transcultural Technology Transfer: Hermeneutics of Technology in Europe, India and China]; Frankfurt u.a.

Petrowski, Henry 1992: To engineers is human: The role of failure in successful design; (1982); New York.

Pool, R. 1997: Beyond Engineering. How Society shapes Technology; New York, Oxford.

Rescher, N. 1983: Risk. A Philosophical Introduction to the Theory of Risk Evaluation an Management; Washington.

Simon, Eberhard 2002: Erhaltung von Technik durch Instandhaltung. Eine technikphilosophische Untersuchung. Masch. Diss. Stuttgart.

Stadler, Gerhard, Anita Kuisle 1999: Hg. Technik zwischen Akzeptanz und Widerstand, Münster u.a.

Tenner Edward 1996: Why things bite back. Technology and the Revenge of Unintended Consequences; New York.

Wagner, Gerald 1992: Vertrauen in Technik. Überlegungen zu einer Voraussetzung alltäglicher Technikverwendung [Trust in Technology: Oberservations to the Presumption of Everyday Technical Uses]; Berlin.

Williams, Rosalind 2002: Retooling. A historian confronts technological change; Cambridge Mass., London.

Winner, Langdon 1992: Autonomous technology. Technics-out-of-control as a theme in political thought; Cambridge Mass. 1992; 1977.

Postphenomenological Inquiry into Brain Research and Human-embodied Mind

Bernhard Irrgang

A long time before authors such as Bruno Latour and Karin Knorr Cetina, Xavier Zubiri in an essay of 1934 pointed out that our modern scientific realization does not imply Ontology in actual senses, but comes from physical coefficients and laws of nature , which the laboratory sciences bring out. This also applies to neuro-sciences. Thus, neuro-philosophy must consider which laboratory sciences it can utilize in its approach. They can only utilize either technology or nature, which is worked on with assistance from laboratory technology. The human mind and human thinking, which are on the action side, can be utilized although not completely in a naturalizing theory. In addition, each laboratory science is dependent on an interpretation of their results (observer perspective). Analytic philosophy based on a methodically limited entrance to the reality ignores a set of neuro-philosophically relevant phenomena, because of using naturalized laboratory methodology it is not able to level the radical positivism in the sciences behind it (Zubiri 1994, P. 54).

I. Naturalization under inclusion of the observer perspective in empirical sciences

The idea of naturalization is completed with new physics. Quantum mechanics represents a radically different program of naturalization than Aristotle (Zubiri 1994, pp. 291-341). The naturalization became a Mathematization. The formal structure of naturalization is spatial temporal. Classical physics, however was concerned with mathematical points. The crisis of modern physics is a consequence of a problem of the ontology of naturalization (Zubiri 1994, pp. 343-351). The quantum theory sets, as they are formulated are, the observational measurements for Technoscience and the human body. Therefore, a form of naturalization needs to be demanded, which includes humans as conscious and science from the observer's perspective. The measuring observer is at the same time the interpreting observer, because fair tests and observations, for example comparisons are not natural processes. The measuring observer was replaced however by positivism like the material naturalization from the own theory formation. Technoscience requires a naturalization under inclusion of the observer's perspective in the empirical sciences.

Zubiris concept of naturalization including the observer's point of view can therefore be connected, with Don Ihde's concept of a non- fundamentalist and justifying Phenomenology. The starting point of my reflexions is the perception-oriented and embodied Intentionality of the observer. As the basis, a kind of instrumental realism (Ihde 1993, P. 12f) is put forward. The substantially unforeseeable and ambivalent relationship between humans and technology, the impossibility, of controlling these relations leads finally to the fact that technologies in their whole are probably more cultures than tools (Ihde 1993, P. 42). Don Ihde's concept of Postphenomenology implies a methodically reflected, empirically oriented phenomenology and technological-culturalistic naturalization for the overcoming of a hidden positivism, both absorbed in natural sciences as well as in analytic philosophy.

II. The observer and perspective: A new paradigm for neurophilosophy

Analytical philosophy is a predominant portion of neuro-philosophical work in the mathematical & physical paradigm of materialistic naturalization. After the failure of strong AI theses the search for an extension of the paradigm of intelligence becomes ever more plausible. Straightforwardly, the phenomenological and hermeneutic philosophy works on the basis of the Technoscience concept on the formulation of a new paradigm, which I would like to call cultural naturalization (whereby mathematical & physical naturalization remains a possible cultural and/or pragmatic & hermeneutic naturalization strategy). The scientific paradigm is limited to the 3PP and became even more of a laboratory science, which can simulate and model brain processes, but human mental states for methodical reasons are not suitable for discussion.

I would like to develop an understanding of the mental states of human-beings, which begin with neuro-philosophy, with the phenomena of mental content, which the interpretation perspective of the 1st Person (1PP) systematically considers. In particular the laboratory sciences do not neglect the methodical entrances. This perspective inevitably changes the results from the now more clear limits of the mathematical and physical modeling in neuro-philosophy and the discovery of the importance of "tacit knowledge" for human mental states, which developed in the execution of human practice. The human mind is to be understood in the following as authority, whereby authority exhibits a plant i.e. an arrangement dimension and an execution dimension. The execution is only to be utilized in the 1PP and in the transcendental-philosophical perspective. From the 3PP it is only indirectly investigated. A philosophy of human-embodied mental state must therefore be ready at the same time from the two perspectives in the sense of a pragmatic „both - and" to be taken, if it is to illuminate the topic appropriately, i.e. philosophically:

We have to take the "both" in a fourfold manor (like a synopsis): (1) The first person perspective (1PP), which makes rather observer-dependent statements ,is about the human mind and its physical-cultural imbedding;

(2) The Second-Person-Perspective (2PP): the discovery of the other human being as someone special as an empirical fact and a transcendental condition of constitution of morality and one-self,

(3) The third person perspective (3PP), those statements about nature, which the human body and brain can make, a scientific reconstruction of the brain processes.

Both dimensions of the question are to be considered in this perspective, whereby from a philosophical perspective the mental state has higher meaning, since also the knowledge of the brain must be finally conceptualized in models in order for it to be understood.

(4) The Lifeworld topic (1PPP) is fundamental for the 1PP as well as for the 3PP of the natural science. Linguistically and technically marked culture is the horizon for human practice and human self understanding as it is for science and laboratory science. Also, brain research is a cultural draft of humans. Storing samples and patterns in the brain, means at the same time a coding of conceptions, memories and experience. These are not things, but execution structures for coding of handling possibilities with sense, meaning and validity. These are not necessarily formal i.e. logical relations itself.

An example: The phenomenon of health/illness splits into the feeling of being ill or healthy and conceptions over health and illness (1PP), into the diagnosis of doctors over the health and illness of a patient i.e. the scientist over normal or deviating measured values and parameters (3PP) and into social dealing with health and illness, such as society's role, institutions like the health service, health insurance and science (1PPP). The three different perspectives cannot be brought into everyday life under one banner, or into the scientific view. Nevertheless .The attempt does not appear to offer no prospects. To -see all three kinds of perspective synoptic together enables us to determine, which is the more precisely based phenomenon.

A new biological & anthropological beginning for the analysis of mental competence with respect to all four perspectives in a philosophical interpretation is looked for in the reconstruction of "tacit knowledge" and "implicit competence" and appropriate fine-motor and linguistic authority. Thereby, a synoptic procedural way is taken as a basis. I understand a methodical procedure in the sense of a repeated „both - and". A consequence of synoptic proceeding is the thought of complementarity of 1PP and the 3PP as a heuristic-methodical manual for a neuro-philosophy in the indication of "both and also". The connection between neural excitation samples and mental meaning is unclear therefore. The relationship between the network of brain and the network of human meanings could be described by a mixture of complementarity, parallelism and Emergence. A clear allocation of brain procedures or entitled and mental processes, does thereby not seem (still) to be manufactured. For a complete understanding of the human mind 1PPP and 2PP are necessary, too.

III. Postphenomenological analyses of human-embodied mind

The mental state is not gained by looking at oneself, and not from reading the human brain, but it develops from the interpretation of mental conditions in other humans, in addition, in understanding about material things like the human culture. I do not want as an infant, what me as an adult would want, I experience my own mentality by processing of the other one, both in personal and in material forms, and in the long run my perspective. My entrance to the mental state is interpretive and re-constructive. The theses from direct entrance to mental state is wrong and the source of various mistakes and errors. Mental state must be able to request obvious neural processes itself, in order to be able to create for itself a basis. The mental state must have its own structure, however this is obviously not independent from neuro-physiological processes. The human mind is generated and embodied in a gradualistic process (Irrgang 2005b).

A human mind examines in neuro-science, another human mind, it is materialization, without being able to really forget thereby that it is human spirit. It is limited to "thing-structures" of the human brain, because it can experimentally utilize nothing different. These „thing-structures" serve to store things (brain processes as noticed ones) or mental things (numbers, words) with which humans can develop operational samples, handling codes and a repeatingability. The handling paradigm of the brain and it's described authority is surely not yet precisely enough described orunderstood to be allowed – it is imperative that further research is carried out here -, but nevertheless an understanding horizon for further research and experimenting is in development. If one materialized going around ability, then one obtains certain arrangements. Now humans have without a doubt a set of arrangements, but some of them are so open with humans that they are no longer called arrangements, but authority.

The straight experiments of brain researchers are theory-loaded. The positive and materialistic naturalization must be limited to the analysis of the brain. For this reason I represent three basic theses:

The materialistic naturalization must be replaced for the formulation of neuroscientific theories such as experiments by a postphenomenological naturalization and a synperspectivic orientation. For a theory of the human mind implicit and explicit aspects of the human spirit and the human soul must be considered. Evolutionary different authority is learned. These must be examined from implicit and explicit perspectives. The language carries the transition from implicit to explicit authority. Not from neural processes and mental contents, but around the „handling authority" from neural processes with mental contents, in particular links with pictures or thoughts around the study.

The human mind as competence has a genesis:

(1) a generic-historical development of the human brain,
(2) an individual development of the human brain,
(3) an individual development of the human mind with implicit and explicit competence and
(4) a history of development of the culture.

IV. The central meaning of different forms of memory

My interpretation suggestion is of a neuro-philosophical kind and orients itself with a regionalization of mental state i.e. the mind of human-beings in the horizon of a philosophical & psychological and mental competence for handling and places the human memory in four fundamental aspects in the center. Learning and re-learning of psychological & mental contents (the content-wise determination of this term requires further interdisciplinary research) like deeply felt situations, dangers, reminded circumstances, linguistic meanings and known facts importantly, concerning the Sensomotoric , survived necessarily and thus indispensably for behavior, thus for the successful manipulation of Sensomotoric and the execution of actions, thus making human practice possible. It is thus quite conceivable and in a certain sense even explainable that psychological & mentality in its basic forms developed through evolution. The social co-ordination with the hunt, when collecting as with storing increases the selection pressure on the self-oriented ability, for understanding (the other one being the situation as itself) to develop further.

The language and the art like technology continue to increase these incentives. Neural activity must work Ever harder and faster, in order to find the incentives for understanding cognition and behavioral patterns. Actions are accompanied by the activation of sensory and motor modules, and thinking actions by reminding and the suggestion of memory modules. We must learn mathematics, entrenched in us as memory samples in the long-term memory, also physics and other sciences must be learned. We use no rules and principles and also learn no formulas by heart, but remind ourselves of appropriate comparable situations. Visually, we deal and shape the formulas. In used situations, we require therefore innate behavioral patterns and also more time. This does not mean that they are steered genetically. But, actually due to learning processes under genetic guidance information is finally processed. Thus, this is the product of genetically innate patterns and learned situations.

All ranges of the human mind co-operate like all brain regions. Neural excitation samples stand for meaning; this is again composed from different characteristics. Mental processes stand for their processing and storage and for the processes of their recalling. Within the evolutionary older parts of the human brain the emotional mind is generated. Hormones and other messenger materials play a central role. Subjectivity is based on mental-state, and mental-ness state results from the interaction of many different factors in the context of brain activity, belonging to different kinds of competencies including tacit knowledge as a competence. It cannot be captured with categories like logical identity or mathematical unit. From a neuro-philosophical view, the human mind arises from senso-motoric competences of cognition and memory. In the course of evolution, different memory modules were added to different forms of cognition, whereby the most fundamental behaviors were steered like the reflexes genetically. Innate behaviors and action patterns can be amazingly flexible:

Stages of mind as competences for handling with memories:

(1) Emotional memory; fixed expirations of behavior and action patterns;
(2) Learning and short time memory, motor as well sensual kind;
(3) Long-term memory, implicitly as well explicit kind;
(4) Linguistic intelligence, abstract memory.

Stage 4 of the modularized memory structure retro-acts like those of the human mind on the lower levels. It gives to these four levels human total cross-linking and total processing. Apart from special modules, like these, only humans exhibit (like special human senso-motoric competence like the skill of human fingers playing a piano and speech modules) the cross-linkings this is totally singular in the animal realm, so that in a biological special way in the evolution of human-beings must be spoken of , which in the long run made the human mind and human culture possible.

V. Observer-centered interdisciplinary neuro-philosophy

Shaun Gallagher and Francisco Varela tell us about the necessity of phenomenology for the cognitive sciences. Both look however rather at the intersubjectivity analysis as well as to the phenomenology of space and time, i.e.

classical phenomenological, while postphenomenology is trying to argue for the observer's point of view. Perspectivity, a synperspectivic perspectivity and the meaning of horizon, i.e. a paradigm for understanding of different forms of practice (in particular about Technoscience) is for the meta-theoretical investigation fruitful. The observer's role and the practice of the realization justified therein of nature and culture imply a philosophy of the human body (Gallagher/Varela, 2001).

In order that neural processes and experiences are identical or correlated, one would have to know how the 1st person perspective and 3rd person perspective compare with one another. There would therefore have to be a point of view, which looks at both perspectives at the same time (Synoptic). That does not seem to work however. The observer's point of view implies the distinction of subject and object. Both around experiments to describe and interpretive developments we need distinctions, so we know that information, which mediates experiments to transform into knowledge, which we can understand. The Interdisciplinary of postphenomenological considerations is the attempt to escape from the pro-crust bed of logical distinctions regarding body-soul and mind-brain problem in the 2,500 years of history of European philosophy. Monism or dualism, identity or interaction, logical terms and firm of terms, valid definitions and ultimate linguistic usage control the philosophical terrain and foundation.

Observation of neural conditions as neural conditions in the 3PP is not experience and perception in the 1PP. However, the 1PP permits only an experience of neural conditions as well as mental conditions, neural conditions such as this is therefore not directly accessible. We have two completely different interpretation paradigms, which are incompatible among themselves. The heuristic postulate of co-relativity of neural processes and mental contents does not seem to be badly occupied empirically. I would like to doubt this co-relativity i.e. parallelism not in principle, ask me however, how this co-relativity can be made philosophically still more plausible, than seems to be the case at present. The fundamental line of interpretation, which I pursue in my philosophy of mental life is that mental life includes a whole number of different authorities and competences. Thus different forms of the mental life and different kinds of neural samples are probably in different areas of the brain to correlate. The naturalistic interpretation proceeds with the impossibility of self realization, from necessary illusions and with self fraud. It remains seized thereby in the paradigm of material naturalization. A philosophical epistemology of mind-brain problem could not be dissolved into a philosophy of the brain.

Rather their starting point will consist of it, with the different methodical perspectives as the starting point of investigation, so that the mind-brain problem i.e. the brain-brain problem cannot be dissolved in the maintained brain-brain relation. In principle, there is a scientific interest to dissolve the 1PP. However "we" are from the 3PP not able to reconstruct mental contents, therefore it is necessary to develop a method, in order to be able to experiment with mental concepts instrumentally. In this case 1 PPP is fundamental. The possibility of an "objective" access to mental stage via brain stage appears to a large extent implausible. A limitlessness of brain, some limitlessness of the knowledge corresponds only indirectly to the reason for this epistemological gap. It is however quite connected with the "fundamental" or "apriori" perspectivity of the observer in the scientific community as a relative apriori, which also makes the distinction in 1PP and 3PP plausible. The observation of mental conditions and the experimental entrance to these are impossible after present knowledge conditions, because they contain a fundamental methodical problem. We know mental conditions directly from experiencing, and knowledge around neural conditions, we must open experimental-scientifically.

The fundamental methodical problem, which made a self realization more difficult for mental life is an epistemological reason, i.e. the impossibility of 1PP and 3PP to be taken together at the same time . The consequences of finiteness and perspectivity of the human brain are impossible. The neural data processing and qualitative content of their experiences are correlated. However, the kind of correlation is today still unknown. Possibly it concerns two different kinds of representation procedures i.e. representation samples in the sense of an ability of a going around competence or skill. Neural representation and sub-symbolic representation in the sense of tacit knowledge prepare the symbolic representation. Red perception and pain experiences are the basis of certain data structures, which are produced by complicated mechanisms.

There is an interior perspective of mental life (1PP) and an external perspective (3PP). They seem to correspond, as far as we know today (how?). There is nevertheless altogether a co-relativity in the large structures, which were worked out in particular by a neuro-philosophy of different forms of human memory. Perhaps it will succeed in the future and seek out and also exhibit further co-relativities in the fine structure of the brain to mental processes. Mindfulness and mental state are either two names for the same thing or complementary brain activities of different natures. Both require a further clarification, which applies also to human consciousness.

The human mind is in accordance with my interpretation, a fortune, ordering ability, an authority, competence, capacity and thus power i.e. authorization to practice. The term arrangement has too many material-end implications, in order to be allowed to considered as particularly suitable. But, it can in an emergency be used as an interpretation assistance to be consulted. The architectonic of the human mind is in the long run evolutionary justified and manifests itself in increasing human authority, for which localized ranges of brain come additionally in the course of evolution. Four fundamental ranges using competence i.e. going around skills of a mental kind can be differentiated:

(1) Competence is first learned going around ability with own sensomotoric skills, the competences of moving and seeing. This leads to implicit knowledge, which was already genetically prepared by dispositions;
(2) The going around ability as the basis for tacit knowledge and being able to learn constitutes a different kind of memory. Now, human abilities and talents can train and develop learning here, an early form of liberty in dealing with oneself and others;
(3) Learning is dealing with language and takes place via its use;
(4) Learning is the use of writing and other formal languages, which includes addition, counting and geometry.

Our subjectivity about the history of the development is due to a cultural-objective framework subjective. Introspections reports are not to be excluded if it concerns neuro-scientific interpretation approaches. The post-phenomenological approach and the hetero-phenomenology as „objectification" of the introspections reports and descriptions of behavior of animals against the background of a theory of the human mind on the basis of the animal people difference is the starting point. On the basis of such a theory, neuro-science cannot expose truth and falsehood as an illusion, since it belongs to the conditions of the experimental arrangement. With positivistic theories we will not solve the mind-brain problem, because we excluded mind from the basic conditions of our experimenting. The mind-brain problem has itself transformed into a problem of the neuro-sciences, neuro-psychology and the intelligence research as well as into a problem of the culture sciences. My view is in particular that philosophy especially for all interdisciplinary research synoptic is responsible.

Abbreviation: 1 PP (First Person Perspective) is the Subjective experience; 3PP (Third Person Perspective) is the Objective experience; 1PPP (First Person Perspective Plural which also includes Cultural aspect) is Inter-subjective experience.

References

Brandom, R. 1994: Making it explicit reasoning, representing and discourse commitment; Cambridge Mass., London.

Corona, N., B. Irrgang 1999: Technik als Geschick? Geschichtsphilosophie der Technik; Dettelbach.

Damasio, A. u. H. 2006: Minding the body; in: Daedalus, Sommer 2006, 15-22.

Dennett, D. 1993: Consciousness explained; 1991, London.

Dreyfus, H. 1985: Die Grenzen künstlicher Intelligenz. Was Computer nicht können (1972);. Königstein/Taunus.

Dreyfus, H. L., St. E. Dreyfus 1987: Künstliche Intelligenz. Von den Grenzen der Denkmaschine und dem Wert der Intuition; Reinbek bei Hamburg.

Gallagher, Sh., F. Varela 2001: Redrawing the Map and Resetting the Time: Phenomenology and the Cognitive Sciences; in: St. Crowell, L. Embree, S. Julian (Hg.). The Reach of Reflection: Issues for Phenomenology's Second Century; 2001.

Ihde, D. 1993: Postphenomenology. Essays in the postmodern context; Evanston.

Ihde, D. 1998: Expanding Hermeneutics. Visualism in Science; Evanston.

Ihde, D., E. Selinger 2003: (Hg.) Chasing Technosciences. Matrix for materiality; Bloomington, Indianapolis.

Irrgang, B. 1992a: Künstliche Intelligenz und Expertensysteme; in: Stimmen der Zeit 210 (1992), 377-388.

Irrgang, B. 1992b: Die Maschinisierung des Subjektes und die rationale Konstruktion der Gesellschaft. Künstliche Intelligenz als Mäeutik eines neuen Bildes vom Menschen und der Art seines Zusammenlebens? in: Joachim Schmidt (Hg.) Denken und denken lassen. Künstliche Intelligenz. Möglichkeiten, Folgen, Herausforderungen; Neuwied, Kriftel, Berlin 1992, 115-154.

Irrgang, B. 1993: - Humanismusstreit um die „Künstliche Intelligenz"; in: Gert Kaiser, Dirk Matejovski, Jutta Fedrowitz (Hgs.); Kultur und Technik im 21. Jahrhundert; Frankfurt/New York 1993, 107-114.

Irrgang, B. 1998: Praktische Ethik aus hermeneutischer Perspektive; Paderborn.

Irrgang, B. 1998b: La Mettries Begründung der Anthropologie; in: J. Beaufort, P. Prechtl (Hg.) Rationalität und Prärationalität. Festschrift für Alfred Schöpf; Würzburg, 81-92.

Irrgang, B. 2001a: Lehrbuch der Evolutionären Erkenntnistheorie; (^{1}1993) München, Basel.

Irrgang, B. 2001b: Technische Kultur. Instrumentelles Verstehen und technisches Handeln; (Philosophie der Technik Bd. 1) Paderborn.

Irrgang, B. 2002a: Technische Praxis. Gestaltungsperspektiven technischer Entwicklung; (Philosophie der Technik Bd. 2) Paderborn.

Irrgang, B. 2002b: Technischer Fortschritt. Legitimitätsprobleme innovativer Technik; (Philosophie der Technik Bd. 3); Paderborn.

Irrgang, Bernhard 2002c: Humangenetik auf dem Weg in eine neue Eugenik von unten? Bad Neuenahr/Ahrweiler.

Irrgang, B. 2003a: Künstliche Menschen? Posthumanität als Kennzeichen einer Anthropologie der hypermodernen Welt?; in Ethica 11/2003/1, 5-33.

Irrgang, B. 2003b: Von der Mendelgenetik zur synthetischen Biologie. Epistemologie der Laboratoriumspraxis Biotechnologie; Technikhermeneutik Bd. 3; Dresden.

Irrgang, B. 2003c: Züchtung als technisches Handeln; in: A. Schäfer, M. Wimmer (Hg.) Machbarkeitsphantasien; Opladen 2003, 67-87.

Irrgang, B. 2004a: Wie unnatürlich ist Doping? Anthropologisch-ethische Reflexionen zur Erlebnis- und Leistungssteigerung; in: C. Pawlenka (Hg.): Sportethik. Regeln, Fairneß, Doping; Paderborn 2004, 279-291.

Irrgang, B. 2004b: Konzepte des impliziten Wissens und die Technikwissenschaften; in: G. Banse, G. Ropohl (Hg.):Wissenskonzepte für die Ingenieurpraxis. Technikwissenschaften zwischen Erkennen und Gestalten; VDI-Report 35; Düsseldorf 2004, 99-112.

Irrgang, B. 2005a: Posthumanes Menschsein? Künstliche Intelligenz, Cyberspace, Roboter, Cyborgs und Designer-Menschen - Anthropologie des künstlichen Menschen im 21. Jahrhundert; Stuttgart.

Irrgang, B. 2005b: Einführung in die Bioethik; München.

Irrgang, B. 2005c: Ethical acts (actions) in robotics; in: Philip Brey, Frances Grodzinsky, Kucas Introna (Hg.) Ethics of New Information Technology. Proceedings of the Sixth International Conference of Computerethics (CEPE 2005); Enschede 2005, 241-250.

Irrgang, B. 2005d: Der Cyborg als der Übermensch Friedrich Nietzsches? Anmerkungen zur Posthumanismusdiskussion; In. R. Kaufmann, H. Ebelt (Hg.) Scientia et Religio. Religionsphilosophische Orientierungen ; Fschr. für Hanna-Barbara Gerl-Falkovitz; Dresden 2005, 315-333.

Irrgang, B. 2007a: Hermeneutische Ethik. Pragmatisch-ethische Orientierung für das Leben in technologisierten Gesellschaften; Darmstadt.

Irrgang, B. 2007b: Gehirn und leiblicher Geist. Phänomenologisch-hermeneutische Philosophie des Geistes, Stuttgart.

Irrgang, B. Klawitter, J. 1990: (Hg.) Künstliche Intelligenz. Stuttgart.

Janich, P. 1992: Grenzen der Naturwissenschaft. Erkennen als Handeln; München.

Libet, B. 2005: Mind time. Wie das Gehirn Bewusstsein produziert, übersetzt von J. Schröder, Frankfurt.

Lenk, H. 2001a: Kleine Philosophie des Gehirns; Darmstadt.

Lenk, H. 2001b: Denken und Handlungsbindung. Mentaler Gehalt und Handlungsregel; Freiburg/München.

Lenk, H. 2004: Bewusstsein als Schemainterpretation. Ein methodologischer Integrationsansatz; Paderborn.

Lenk, H. 2005: Emotionen und Gefühle werden schemainterpretatorisch erfasst, sind aber biologisch evolutionär verankert; in: M. Wimmer, L. Ciompi (Hg.): Emotion – Kognition – Evolution. Biologische, psychologische, soziodynamische und philosophische Aspekte; O.O. 2005, 247-272.

Markowitsch, H. 2002: Dem Gedächtnis auf der Spur. Von Erinnern und Vergessen; Darmstadt.

Metzinger, Th. 1993: Subjekt und Selbstmodell. Die Perspektivität phänomenalen Bewusstseins vor dem Hintergrund einer naturalistischen Theorie mentaler Repräsentation; Paderborn u.a.

Northoff, G. 2000: Das Gehirn. Eine neurophilosophische Bestandsaufnahme; Regensburg.

Pauen, M., G. Roth 2001: Neurowissenschaften und Philosophie. Eine Einführung, München.

Polanyi, M. 1998: Personal knowledge. Towards a Post-Critical Philosophy; London, 1958.

Walther, H. 1999: Neurophilosophie der Willensfreiheit. Von libertarischen Illusionen zum Konzept natürlicher Autonomie (1997), Paderborn.

Zimmerli, W. 1996: Zeit als Zukunft. Die menschliche Konstruktion der Zeit. Rhythmen und Uhren, Cyber-Medien-Fiktion und Technikfolgenabschätzung. Vom Handeln im Mensch- Maschine-Tandem; in: K. Weis (Hg.): Was ist Zeit? Teil 2: Entwicklung und Herrschaft der Zeit in Wissenschaft, Technik und Religion, München 1996, 221-248.

Zimmerli, W.; St. Wolf 1994: Künstliche Intelligenz. Philosophische Probleme; Stuttgart.

Zimmerli, Walther 2002: Jenseits von Zähmung oder Züchtung – die Ablösung der künstlichen Intelligenz durch den Netzwerkmenschen; in: K. Kegler, M. Kerner (Hg.): Der künstliche Mensch. Körper und Intelligenz im Zeitalter ihrer technischen Reproduzierbarkeit; Köln u. a. 2002, 75-103.

Zubiri, X. 1994: Naturaliza, Historia, Dios; (1944) Madrid.

Visions of Technology

Bernhard Irrgang

In the 20th century, with the advancement of technology as a world-historical power, philosophers have spoken of „the end of History"(Arnold Gehlen) and „the end of Philosophy"(Martin Heidegger). Technological progress replaces philosophy and reflection and thereby accompanies the end of cosmologically oriented, i.e. Metaphysics. Although it focused itself on the ecological crisis for a short time at the end of 20th century. The 20th century was a revival of nature-philosophical thinking. Also, the cultural unity of a technological kind was threatened and is still threatened at the current time. But the current trends are moving in an opposite direction, and offering new meaningful thinking about culturally embedded technology – which in turn refers to the cultural understanding of technological development.

Since the Industrial revolution, the idea of alternative technological futures has become increasingly central to plans for technical decisions. Thus arises the more general question of the concept for the future of technology, which we want to conceptualise in our vision, and for that purpose a technical utopia, perhaps a technological world-view, is necessary. Technological development in its ambivalent form and the future of technological development replaces the paradigm of the technological progress. Also the generic future of human beings, to which technological progress has been directed since the Enlightenment, is an insufficiently broad concept. It needs to be changed and integrated into a concept of Sustainable Development. The concrete formation of human beings in its Bodily Existence[21] should be placed at the center of the evaluation of technological progress. Since the Industrial Revolution, the coincidental technological evolution with its acceleration effects is taken over by an organizational model of projected technologies.

Most organizational theories and traditional forms of ethics want to limit the borders of technologies, but without transforming the technological development. The formulation of moral limits, which propagates the return to a new simplicity, is the usual approach. However, one must acknowledge the limits of moral and political fixing of boundaries. The western theory means, alternative forms of the technologization (in other cultures and societies) are not possible.

---

21 Irrgang notion of bodily existence (Leiblichkeit) (See Irrgang, B. 2003: Kuenstliche Menschen? Posthumanitaet als Kennzeichen einer Anthropologie der hypermodernen Welt? [Artificial humans? Posthumanity as the Expression of anthropology of the Hypermodern world?]; in EHICA 11/2003/1, pp. 5-33).

With a culturalistic theory in the background, the idea of an alternative technological future can be developed. The economical costs of regularization have to be considered and the conditions of the dominant economic culture have to be questioned. Technology is always actually adapted to changing conditions, therefore alternative technologies are possible.

The compensation theory of the human sciences (Geisteswissenschaften) by Odo Marquard goes back to the philosophical studies of Hermann Luebbe. Besides, Joachim Ritter perceives an addition of the scientific & technological development by the Geisteswissenschaften. In particular, it maintains a compensation of history by technology studies and natural sciences. However, preceding the hypothesis of compensation, an inherent thinking from the assumption is needed; the substantiality of previous lifeforms could regain the liability with the assistance of Geisteswissenschaften (human sciences).

The hypothesis of compensation keeps the myth of the two cultures intact. During this process, the compensation of adversity of modernisation is publicized and the complementary function of the historical culture sciences is ignored. The hypothesis of compensation refers only to the part of the natural scienceswhich produces technical-industrially used knowledge. Joachim Ritter in his studygives a functionalistic analysis for the authorization of existence of the human sciences (Geisteswissenschaften) by modeling the hypothesis of compensation.

Natural sciences and technology are embedded into a network of tradition. Innovations, in return are subsequently bounded with transformation of tradition. Technological development can be understood as a cultural-historical process. Needs and value conceptions resist technological development, whereas cultural perspectives are more important than general subordinates. This kind of resistance – however preferentially made by philosophers – is not analytical, although it would not be uninteresting for ethical evaluations. Also, an ethics of the technological development is not to be understood as compensation. The thesis of dealing with technological knowledge and action implies also another concept of ethics. At this junction, ethics is not added from the outside for technological development, but from the very beginning ethical evaluation/assessment is a part of technological action. This also changes the concepts of modernization. Technological development does not only happen exclusively for its own sake, even if this has sometimes appeared this way from the ground. Therefore, a new concept of modernization is urgently needed on the national and global scale.

Cultural models are criticising the particular technological alternatives as inhuman or ecologically harmful and focus on adapted or intelligent solutions. Ideas of naturalness or humanity have always been included into a path-dependent orientation of particular technological developments. The substantial paths of individual technology advancement result from an interaction of various selected and limited conditions. With the dynamic of variation and construction, particular fields of technological development, routines of construction and paradigmatic solutions have been worked out. The routines of construction are established in the „State of Technology". In this respect a path – dependency of technological developments results from the practice of technology advancement. There is no central authority which would structure the entire development. Therefore, one has to take various contexts of structuring into consideration if researchers want to structure a particular frame of conditions.

Technological action is defined as dealing with technological practice which goes along with traditions, paths of development and selective models. Models can be worked out in view of development trends. Here one can speak of trends, but not of preformed technological ideas. Though construction patterns of a technological kind seems to be more than mere social constructions of technological developments. Technology does not arise from one single consistent project, so it cannot be planned in advance, but is being developed out of a gradual constitution process. Yet neither Technology Assessment (Technikfolgenabschätzung) nor Technology Structuring can be done only by one single project in advance but also has to be taken as a gradual process of constitution and reflection. Nevertheless the openness of technological development has nothing to do with irrationality, one can realize it and in the light of this openness one can also act rationally.

In accordance with the culturalistic turn, technological development is taken as a model for cultural development. The fitness of certain means for certain purposes can always be estimated by success and failure, i.e. by reaching and missing the certain purpose (goal) or by certain means to judge. The success of technological action does not have to depend on the actual agreement of various groups, but any time it must demonstrate transculturally and prove its practical worth. Technical know-how has not only the characteristics of continuation and unreversibility, but it is also not revisable, which shows clearly a cumulative character.

In this sense, in principle the technology goes through a progress of direction. Technological progress, as a progress of cognition and realization, can be methodically restored as a hierarchically structured and diversifying acting competence. This always enriches to become available for action. Philosophy reconstructs those social practices, which imply poietic instrumental action (poesis action) or the result of a poietic action: such as practices of technological development and technological production, practical use of technology and the practices in which technology must be removed from the context of usage (disposal etc.). Therefore the main feature of these artefacts is their relict character.

A large number of decisions are needed to ensure the technical organization runs smoothly, efficiently and inconspicuously; but only relatively few organizations let questions about the technical organization appear as a precarious conflict-laden business. The large and inhomogenous group of the Technics Designer is to be structured in such a way that it itself works in principle, for which relevant differences bring ethical questions into discussion. Technology organising individuals usually received no professional training in ethics. Different approaches of business ethics find entrance (an easy access) into an operational practice are also increasingly being introduced in industrialized countries. But compared to this, the application of ethical reflection for entrepreneurial technological organization is still in a very early phase.

Technology organising practical men stand with their needs to receive in concrete individual cases, are also continuously seeking ethical assistance outside of the universitaeren system. Most people who participate in technological organization do not have the information about those different aspects, whereby the possible ethical consultation is already prevented, which took part on the other hand. In each case the contact with rather highly deterring expenditure is connected. Ethical criteria are an integral component of the new modernization discussion.

Technological progress lies first within the instrumental range and requires a pragmatic & utilitarististic justification. In addition, progress in technological action has an ethical dimension. To that extent, pragmatic and ethical legitimacy of examination can be differentiated methodically, but cannot be completely separated. The legitimacy regarding a practice must take the limitations of tradition and teachableness into account. The legitimacy can only be achieved regarding the uncertainty of the technology consequences and future development. Overall a deficit of complexity theory and the unsatisfactory causality of prognosis make the estimation more difficult for future developments, which however are not completely impossible.

A part of the acceptance crisis of modern technology would have to be attributed to missing visions of technological development. People talk rather about the frightful visions than about positive utopias, as they are significant for technological development. During the earlier stages of technological development optimistic progress was allowed. We are in the age of rapid technological change, in which however individual innovations, trend and megatrend are the characteristics. On the Way towards the High Technology Civilization, it gives a whole number of successes within the range of automobiles of the future, power supply, High-tech medicine are new building materials. What is missing is a model for cross-linking and embedding of these new technologies, a vision for a coherent cultural paradigm, which, finally helps to prepare the acceptance. This spells a new category of modernization, which has a new meaning.

A vision of technology is not to be confounded and exchanged with prognosis. They have not any prognostic features. It would not like to point out, what would be likely to occur in the future. But, it tries a structuring of individual fields of technology, a security of technology and its cross-linking, not least in the indication a succeeding technical progress i.e. technological practice. However, a practice based on the knowledge of its basic conditions and its background justifications. Prosperity and use were long time legitimizing horizons for technological practice. Today at a rudimentary level, at least the succeeding life and human practice, is now morally, pragmatically, strategically or instrumentally and technologically, judged by other criteria. This is a starting point for a better human practice. At the same time, it is a challenge for technicians who are responsible for the progress in the technological development, who believe that technologically working order and fulfilment is already sufficient. I would not like to deny, the success and failure of human practice in the central indicators for the evaluation of the technological practice.

Technical criteria alone are not sufficient, but cultural-civilization models with a moral component, such as sustainability or long-term responsibility will become part of a technology reflection culture. Culture of the technological reflection, in exchange for energisation, in order to raise the question of acceptance between the engaging technicians. On the one hand, engineers and economical transaction need to be reflected and on the other hand, specialists for the acceptability of technologies are needed. Communication, mobility and knowledge regarding information must be brought together in harmony with ecological, civilization and communicative embedding paradigms, in order to be able to finally clarify the questions of acceptability of technological practice. These issues must be examined in a transdisciplinary research practice (Irrgang 2002a; Irrgang 2002b; Irrgang 2006).

Epilogue: Questions Concerning Technology

Peter-Paul Verbeek

German philosophy of technology has a long tradition. Or, better, the philosophy of technology has a long German tradition. For some, this tradition might seem to be dominated by the work of Martin Heidegger. But even though Heidegger's 1953 Die Frage nach der Technik (The Question concerning Technology) is still an icon for German philosophy of technology, the tradition encompasses much more. Many philosophers of technology consider the appearance of Ernst Kapp's 1877 Grundlinien einer Philosophie der Technik (Fundamentals of a philosophy of technics) as the birth certificate of the field. Since then, many other German philosophers have given shape to a broad variety of approaches and theories of technology. Kapp's dialectical approach was taken up by Arnold Gehlen, who firmly established a philosophical-anthropological approach to technology. And after Gehlen, a new generation of philosophers entered the field, like Hans Lenk, Friedrich Rapp, Günter Ropohl, und Walther Zimmerli, who brought in a broad variety of perspectives, including more internally oriented engineering perspectives, and influences from analytic philosophy.

Firmly rooted in this rich tradition, the work of Bernhard Irrgang initiates an interesting new development. Characteristic of his approach is a central focus on developing a 'philosophical anthropology of technology', on the basis of which he addresses a broad variety of issues and questions, ranging from the ethics of robotics to biotechnology, and from technology transfer to issues of trust. His work shows a deep engagement with the social roles and impacts of technologies. But rather than taking on a technophobic character, his work also engages with technological practices and opens possibilities to give these practices a good shape.

Also in the international context of philosophy of technology this is an innovative and much awaited approach. Against the current focus on applied ethics, Irrgang's work has the power to bring back the philosophy and ethics of technology to the bigger and more encompassing questions that are at its roots. Irrgang's work opens new perspectives on the impact of technology on the future of humanity and human culture, and on the ethical questions related to this impact.

Both Irrgang's work and the philosophy of technology, therefore, deserve this first translation of some of his work into English, carefully rendered by Arun Kumar Tripathi. Disclosing these highly original and illuminating texts to an Anglophone audience is a great enrichment for the field. One can only hope that this will be the beginning of many more Questions Concerning Technology to come, which German philosophy of technology will both raise and answer in its own, sophisticated way.

Appendix: Bibliography of Bernhard Irrgang's philosophical Works

I. Monographies

- Skepsis in der Aufklärung. Zur Rekonstruktion der Bedeutung skeptischer Argumentation und ihrer Widerlegung in Versuchen der Rechtfertigung ihres Anspruchs als 'Siècle philosophique'; 472 S., Frankfurt 1982.

- Christliche Umweltethik. Eine Einführung; 351 S., München, Basel 1992, UTB-Band 1671.

- Lehrbuch der Evolutionären Erkenntnistheorie; 359 S., München, Basel ^{1}1993; UTB-Band 1765; völlig überarbeitete und erweiterte Neuauflage 2001.

- Grundriss der medizinischen Ethik; 295 S., München, Basel 1995; UTB-Band 1821 (Übersetzung mit neuem Vorwort ins Japanische 2002).

- Forschungsethik Gentechnik und neue Biotechnologie. Grundlegung unter besonderer Berücksichtigung von gentechnologischen Projekten an Pflanzen, Tieren und Mikroorganismen; 411 S., Stuttgart 1997.

- Praktische Ethik aus hermeneutischer Perspektive[Practical Ethics from a Hermeneutic Perspective]; 266 S., Paderborn 1998; UTB-Band 2020.

- Technische Kultur. Instrumentelles Verstehen und technisches Handeln [Technological Culture. Instrumental Understanding and Technical Action]; (Philosophie der Technik Bd. 1) 240 S. Paderborn 2001.

- Technische Praxis. Gestaltungsperspektiven technischer Entwicklung[Technological Practice. Design Perspectives of Technical Development]; (Philosophie der Technik Bd. 2); 238 S.; Paderborn 2002.

- Technischer Fortschritt. Legitimitätsprobleme innovativer Technik [Technological Progress. Legitimacy Problem in Innovative Technology]; (Philosophie der Technik Bd. 3); 218 S.; Paderborn 2002.

- Natur als Ressource, Konsumgesellschaft und Langzeitverantwortung. Zur Philosophie nachhaltiger Entwicklung [Nature as Resource. Consumer Society and Long time Responsibility. Towards a Philosophy of Sustainable Development]; 259 S.; Technikhermeneutik Bd. 2; Dresden 2002.

- Humangenetik auf dem Weg in eine neue Eugenik von unten? 122 S., Bad Neuenahr/Ahrweiler 2002.

- Von der Mendelgenetik zur synthetischen Biologie. Epistemologie der Laboratoriumspraxis Biotechnologie [From the Mendel's Genetics to Synthetic Biology. Epistemology of Laboratory Practice in Biotechnology]; Technikhermeneutik Bd. 3; Dresden 2003.

- Posthumanes Menschsein? Künstliche Intelligenz, Cyberspace, Roboter, Cyborgs und Designer-Menschen - Anthropologie des künstlichen Menschen im 21. Jahrhundert [Posthuman bodily existence? Artificial intelligence, Cyberspace, Robots, Cyborgs and Designer babies – Anthropology of Artificial Humans in 21st Cenntury]; Stuttgart 2005.

- Einführung in die Bioethik [Introduction to Bioethics]; München 2005.

- Technologietransfer transkulturell. Komparative Hermeneutik von Technik in Europa, Indien und China [Transcultural technology transfer. Comparative Hermeneutics of Technology in Europe, India and China]; Dresden Philosophy of Technology Studies 1; Frankfurt u.a. 2006.

- Hermeneutische Ethik. Pragmatisch-ethische Orientierung in technologisierten Gesellschaften [Hermeneutics Ethics. Pragmatic-ethical Orientation in Technological Society]; Darmstadt 2007.

- Gehirn und leiblicher Geist. Phänomenologisch-hermeneutische Philosophie des Geistes [Brain and Embodied Mind. Phenomenology and Hermeneutic Philosophy of Mind]; Stuttgart 2007.

- Technik als Macht. Versuche über politische Technologie [Technology as Power. Towards a Politics of Technology]; Hamburg 2007.

- Philosophie der Technik [Philosophy of Technics and Technology]; Darmstadt 2008.

- Der Leib des Menschen. Grundriss einer phänomenologisch-hermeneutischen Anthropologie [Body of Humans. Towards a Phenomenology and Hermeneutics Anthropology]; Stuttgart 2009.

- Grundriss der Technikphilosophie: Hermeneutisch-phänomenologische Perspektiven [Compendium of the Philsophy of Technology. Hermeneutics-phenomenological Perspectives]; Würzburg, 2009.

II. Series editor

Technikhermeneutik; Dresden (since 2002)

Dresden Philosophy of Technology Studies/Dresdner Studien zur Philosophie der Technologie; Frankfurt, Berlin, Bern, Bruxelles, New York, Oxford, Wien (since 2005)

III. Joint Authorship

- zusammen mit Nestor Corona; Technik als Geschick? Geschichtsphilosophie der Technik; 271 S.; Dettelbach 1999.

- C. R. Bartram, J.P. Beckmann, F. Breyer. G. Frey, C. Fonatsch, B. Irrgang, J. Taupitz, K.-M. Seel, F. Thiele; Humangenetische Diagnostik. Wissenschaftliche Grundlagen und gesellschaftliche Konsequenzen; Berlin et al. 2000.

- zusammen mit: Michael Göttfert, Matthias Kunz, Joachim Lege, Gerhard Rödel, Ines Vondran; Gentechnik in der Pflanzenzucht. Eine interdisziplinäre Studie; Dettelbach 2000.

- Die deutschen Bischöfe (Kommission für gesellschaftliche und soziale Fragen, Kommission Weltkirche) Nr. 29 Der Klimawandel: Brennpunkt globaler, intergenerationeller und ökologischer Gerechtigkeit. Ein Expertentext zur Herausforderung des globalen Klimawandels; Bonn September 2006.

- Hubig, Christoph, Hans Poser 2007: (Hg.) Technik und Interkulturalität. Probleme, Grundbegriffe, Lösungskriterien; VDI-Report 36; Düsseldorf.

IV. Joint Editorship (Books)

- mit Hans Michael Baumgartner; Am Ende der Neuzeit? Die Forderung eines fundamentalen Wertwandels und ihre Probleme; 205 S., Würzburg 1985 (eigene Beiträge zur Neuzeitproblematik, zur Spätphilosophie Husserls und zur Genese der Wertphilosophie).

- mit Jörg Klawitter und Klaus Philipp Seif; Wege aus der Umweltkrise. Dokumentation einer Tagung der Katholischen Akademie Rabanus Maurus, Wiesbaden, und der Studiengruppe Entwicklungsprobleme der Industriegesellschaft (STEIG) e.V., Würzburg, 199 S., Frankfurt, München 1987 (eigener Beitrag zur Begründungsproblematik der Umweltethik).

- mit Matthias Lutz-Bachmann; Begründung von Ethik. Beiträge zur philosophischen Ethik-Diskussion heute, Würzburg 1990 (eigener Beitrag zur Evolutionären Ethik).

- mit Jörg Klawitter; Künstliche Intelligenz, 163 S., Stuttgart 1990 (gemeinsame Einleitung mit J. Klawitter).

- mit Hans Peter Böhm, Helmut Gebauer; Nachhaltigkeit als Leitbild für Technikgestaltung; Forum für Interdisziplinäre Forschung 14 Dettelbach 1995 (Beitrag zur Technikgestaltung).

- mit Ricardo Maliandi; Technikphilosophie in Lateinamerika. Themen, Probleme und Entwicklungsperspektiven am Beginn des 21. Jahrhunderts; 216 S.; Technikhermeneutik Bd. 1; Dresden 2003.

- mit Sybille Winter: Modernität und kulturelle Identität. Konkretisierungen transkultureller Technikhermeneutik im südlichen Lateinamerika; Frankfurt u.a. 2007.

V. Papers on the Ethical Discussion

- (Mehrverfasserbeitrag) Die Interdependenz von Ethik, Ökonomie und Politik in der Analyse der Industriegesellschaft aus philosophischer Sicht; in: Reiner Kümmel/Monika Suhrcke (Hg.); Energie und Gerechtigkeit, München 1984, 137-151.

- AIDS aus ethisch-theologischer Perspektive, in: Das öffentliche Gesundheitswesen 50 (1988), 197-201.

- Das Ethische als bloße Funktion des Religiösen? Eine Auseinandersetzung mit Eugen Drewermanns Interpretation der J-Urgeschichte; in: Münchener Theologische Zeitschrift 39 (1988), 139-143.

- Sittliche Orientierung der Grundlagenforschung?; in: Hirschberg 42 (1989), 371-377.

- Ethische Implikationen globaler Energieversorgung; in: Stimmen der Zeit 207 (1989), 607-620.

- Mensch und Mitwelt in der neueren katholischen Schöpfungstheologie und Umweltethik; in: Tilman Evers (Hg.); Schöpfung als Rechtssubjekt? Schriftenreihe der Evangelischen Akademie Hofgeismar 1990, 80-102.

- Naturrecht als Entscheidungshilfe? Am Beispiel der Bewertung gentechnischer Verfahren aus ethisch-theologischer Perspektive; in: Marianne Heimbach-Steins (Hg.); Naturrecht im ethischen Diskurs; Münster 1990, 67-98.

- Das Konzept des Regelkonsequentialismus als Grundlegung einer Wirtschaftsethik; in: Michael Wörz, Paul Dingwerth, Rainer Öhlschläger (Hg.); Moral als Kapital. Perspektiven des Dialogs zwischen Wirtschaft und Ethik; Stuttgart 1990, 235-252.

- Stichwort: Ethik; in: Arzt und Christ 36 (1990), 136-138.

- Solidarität mit der Natur? Eine Ortsbestimmung umweltethischen Denkens; in: Jörg Klawitter, Reiner Kümmel, Gerhard Maier-Rigaud (Hg); Natur und Industriegesellschaft, Berlin, Frankfurt, New York 1990, 91-111.

- Hat die Natur ein Eigenrecht auf Existenz? Anmerkungen zur Umweltethik-Diskussion; in: Philosophisches Jahrbuch 97 (1990), 327-339.

- Sittliche Bewertungs-Kriterien der Human-Gentechnik; in: Stimmen der Zeit 209 (1991), 239-253.

- Transgene Tiermodelle im biomedizinischen Experiment. Forschungsethische Überlegungen; in: Forum für interdisziplinäre Forschung 4 (1991), Heft 1, 54-66.

- Naturrechtliche Begründung der Umweltethik?; in: „Aus Politik und Zeitgeschichte", Beilage zur Wochenzeitung DAS PARLAMENT, B 33/1991, 39-46.

- Grundlagen der Wirtschaftsethik; in. M. Lutz-Bachmann (Hg.); Freiheit und Verantwortung. Ethisch handeln in den Krisen der Gegenwart; Berlin 1991, 187-213.

- Verantwortete Forschungsfreiheit bei der Anwendung der Gentechnik; in: Ethik und Sozialwissenschaften 2(1991)4, 595-597.

- Auf dem Weg zu einem neuen Weltethos? Kirchliche und ökumenische Verlautbarungen zur Umweltproblematik; in: EB 38 (1992) 93-96.119-120.

- Eugen Drewermann und die theologische Ethik; in: Bernadette Benedikt, Alfred Sobel (Hg.): Der Streit um Eugen Drewermann, Wiesbaden. Berlin 1992, 123-139.

- Mensch und Umwelt. Die Verantwortung für die Schöpfung; in: Schulreport 1992/5/6, 17-18.

- mit Matthias Kunz: Krankenhauspsychiatrie und Ethik; in: Forum für interdisziplinäre Forschung 10 (1992), 103-120.

- Teleologie der Natur. Zur Begründung christlicher Umweltethik; in: Michael Schramm, Udo Zelinka (Hgs.): Um des Menschen willen. Moral und Spiritualität. Fschr. f. Bernhard Fraling; Würzburg 1994, 231-251.

- Gerechtigkeit als Grundlage einer internationalen Umweltpolitik; in: Sozialwissenschaftliche Informationen 23 (1994), I, 40-49.

- Defensivmedizin und Zwangsbehandlung in der Krankenhauspsychiatrie aus der Sicht der Patientenautonomie; in: Ethica 3(1995),1, 71-74.

- Ein Ethos ökologisch orientierter Humanität als Weltethos; in: Ekkehard Kessler (Hg.): Ökologisches Weltethos im Dialog der Kulturen und Religionen, Darmstadt 1996, 216-225.

- Zwischen Paternalismus und Patientenautonomie. Ein neues Paradigma setzt sich in der Medizinethik durch; in: Philosophischer Literaturanzeiger 49 (1996), 80-91.

- Das Problem umweltethischer Zielsetzung; in: B. Köstner, M. Vogt (Hgs.): Das Problem umweltethischer Zielsetzung; Dettelbach 1996, 71-84.

- Am Ende der Anthropozentrik? Wie lassen sich unsere Verpflichtungen gegenüber der Natur begründen?; in: Gotthard Fuchs, Guido Knörzer (Hg.) Tier, Gott, Mensch. Beschädigte Beziehungen; Frankfurt 1998, 13-32.

- Wozu können Klonierungsverfahren dienen: ethische Bewertungskriterien; in: Johannes S. Ach, Gerd Brudermüller, Christa Runtenberg (Hgs.): Hello Dolly? Über das Klonen; Frankfurt/M. 1998, 72-89.

- Ethische Aspekte der Gentechnik; in: Kursbuch Umwelt. Journal des Sächsischen Staatsministeriums für Umwelt und Landwirtschaft 1/98, 30-31.

- Criterios de evaluación moral de la tecnologia genética humana; in: Cuadernos de Etica 23/24 1997 (erschienen Okt. 1998), 83-103.

- Hermeneutische Ethik zwischen Logik der Kooperation und Autonomie des Sittlichen; in: Hans Michael Baumgartner, Winfried Böhm, Martin Lindauer (Hgs.): Streitsache Mensch. Zur Auseinandersetzung zwischen Natur und Geisteswissenschaften; Stuttgart 1999, 189-210.

- Globalisierung der technologisch-ökonomischen Entwicklung und die Wiederkehr des Verantwortungssubjektes; in: Hans-Günter Gruber, Benedikta Hintersberger (Hgs.) Das Wagnis der Freiheit. Theologische Ethik im interdisziplinären Gespräch. Johannes Gründel zum 70. Geburtstag; Würzburg 1999, 343-353.

- Gemeinwohl geht vor Eigennutz. Eine Auseinandersetzung mit dem Kommunitarismus; in: P. Fonk, U. Zelinka (Hgs.): Orientierung in pluraler Gesellschaft. Ethische Perspektiven an der Zeitenschwelle. Festschrift zum 70. Geburtstag von Bernhard Fraling; Freiburg, Freiburg, Wien 1999, 149-164.

- Gentechnik in der Tierzüchtung; in: Thomas Hausmanninger, Scheule, Rupert (Hg.) ... geklont am 8. Schöpfungstag. Gentechnologie im interdisziplinären Gespräch; Augsburg 1999, 227-242.

- Plädoyer für die Europäische Bioethik-Konvention; in: LAGH (Hg.) Perspektiven 5 (1999), 99-105.

- Prädiktive Medizin und Behinderung; in: Wissenschaftliche Zeitschrift der TU Dresden 48 (1999) Heft 5/6, 13f.

- Keine überzeugende evolutionäre „Genealogie der Moral"; Replik zu G. Dux: Historisch-genetische Theorie der Moral. Die Moral im Schisma der Logiken; Ethik und Sozialwissenschaften 11/2000, 1, 30f.

- Hermeneutik und Ethik; in: Ethica 8 (2000), 3, 267-278.

- Der Krankheitsbegriff der prädiktiven Medizin und die humangenetische Beratung; in: A. M. Raem, R.W. Braun, H. Fenger, W. Michaelis, S. Nikol, S.F. Winter (Hg.) Gen-Medizin. Eine Bestandsaufnahme; Berlin et al. 2000, 651-660.

- Etica y Hermeneutica; in: Excritos de Filosofia (Buenos Aires) 39/40 2001, 3-26.

- Das Stichwort: Hermeneutische Ethik; in: Information Philosophie 2/2002, 50-52.

- (W. E. Thasler et al.): Die Verwendung menschlichen Gewebes in der Forschung. Ethische und rechtliche Gesichtspunkte; in: Deutsche Medizinische Wochenschrift 2002:127, 1397-1400.

- Ethical Issues of Genetic Manipulation of Lifestock; in: ISAG: XXVIII International Conference of Animal Genetics 11.-15.8.2002; Göttingen, 21-26.

- Institutionalisierung praktischer Ethik. Ethiktransfer, Ethikimplementation oder angewandte Ethik; in : Ethica 11-2003.1, 3.

- Hermeneutische Ethik; in: Zeit-Schrift für interdisziplinäre Bildung und praktische Philosophie 2004, 106-113.

- /Frank Oehmichen: Ethische Fragen der künstlichen Ernährung; in: Ethica 13 (2005) I, 69-91.

- Grundlagen für ein neues Bild vom Lebendigen; Kritik des Hauptartikels : Die Entschlüsselung des Humangenoms –ambivalente Auswirkungen auf Gesellschaft und Wissenschaft; in : Erwägen. Wissen. Ethik (EWE) 16 (2005) –2, 181-183.

- Ethical acts (actions) in robotics; in: Philip Brey, Frances Grodzinsky, Kucas Introna (Hg.) Ethics of New Information Technology. Proceedings of the Sixth International Conference of Computerethics (CEPE 2005); Enschede 2005, 241-250.

- Hermeneutische Ethik; übersetzt von Nohihiro Yokochi in: Moralia Nr 13, The Association of Ethical Studies Tohoku University, Sendai, Japan 2006, 54-67.

- Ethical Problems of the Embryo and Stem Cell Research; Koreanische Akademie der Wissenschaften 3/2006; http://www.aks.ac/aks kor/upload/Rsch/stemcellkorea.doc.

- Ethische Bewertung der Energieerzeugung aus Biomasse; in Forum TTN 17 (Mai 2007), 3-17.

- Realisierbarkeit sittlicher Urteile als ethisches Kriterium – Implikationen für Theorien angewandter Ethik; In: M. Zichy, H. Grimm (Hg.) Praxis in der Ethik. Zur Methodenreflexion in der anwendungsorientierten Moralphilosophie; Berlin, New York 2008, 359-386.

VI. Papers on the Philosophy of Technology

- Leitlinien einer Ethik der Gentechnik. Vorüberlegungen zu einer Ethik der Biotechnologie; in: Naturwissenschaften 77 (1990), 569-577.

- Zum Ansatz einer Forschungs- und Standesethik für die Gentechnik; in: Hans Lenk, Matthias Maring (Hg.); Technikverantwortung. Güterabwägung, Risikobewertung, Verhaltenskodizes; Frankfurt, New York 1991, 263-284.

- Sittliche Fragestellungen und Grenzziehungen bei der Anwendung der Biotechnologie in der Tierproduktion; in: Züchtungskunde 63 (3); 1991, 247-257.

- mit Stephan Feldhaus; Ethik der Energieversorgung. Kriterien, Maximen, Handlungsregeln; in: Forum für interdisziplinäre Forschung 4 (1991), Heft 2, 1-13.

- Künstliche Intelligenz und Expertensysteme; in: Stimmen der Zeit 210 (1992), 377-388.

- Die Maschinisierung des Subjektes und die rationale Konstruktion der Gesellschaft. Künstliche Intelligenz als Mäeutik eines neuen Bildes vom Menschen und der Art seines Zusammenlebens? in: Joachim Schmidt (Hg.) Denken und denken lassen. Künstliche Intelligenz. Möglichkeiten, Folgen, Herausforderungen; Neuwied, Kriftel, Berlin 1992, 115-154.

- Ethische Aspekte der Biotechnik; in: H. Wilhelm Schaumann-Stiftung (Hg.): Biologisch-technische Entwicklung in der Tierproduktion (14. Hülsenberger Gespräche); Hamburg 1992, 36-46.

- Humanismusstreit um die „Künstliche Intelligenz"; in: Gert Kaiser, Dirk Matejovski, Jutta Fedrowitz (Hgs.); Kultur und Technik im 21. Jahrhundert; Frankfurt/New York 1993, 107-114.

- Gentransfer bei Tierversuchen und in der Tierzucht. Ethische und juristische Grenzen; in: Politische Studien 44 (1993) 332, 76-81.

- Dimensionen des Verantwortungsbegriffes in der Technologie-Zivilisation; in: Ethica 2 (1994) H. 2, 155-169.

- Verantwortungsethik in der technischen Zivilisation; in: M. Heimbach-Steins, A. Lienkamp, J. Wiemeyer (Hgs): Brennpunkt Sozialethik. Theorien, Aufgaben, Methoden. Fschr. für Franz Furger; Freiburg, Basel, Wien 1995, 403-417.

- Von der Technologiefolgenabschätzung zur Technikgeneseforschung. Leitbilder in einer verantwortungsethischen Konzeption der Technologiegestaltung; in: H.P. Böhm, H. Gebauer, B. Irrgang (Hgs.) Nachhaltigkeit als Leitbild für Technikgestaltung; Forum für Interdisziplinäre Forschung 14 (1995), 13-25.

- Von der Technologiefolgenabschätzung zur Technologiegestaltung. Plädoyer für eine Technikhermeneutik; in: Jahrbuch für Christliche Sozialwissenschaften 37 (1996), 51-66

- Die ethische Dimension des Nachhaltigkeitskonzeptes in der Umweltpolitik: in: Ethica 4 (1996) H. 3, 245-264.

- Technik als Wachstumsmotor? Wie und warum wächst Technik?; in: Evangelische Akademie Bad Boll (Hg.) Wachstum. Über die Wurzeln der Dynamik von Technik und Wirtschaft; Bad Boll 1997.

- Nachhaltigkeit als Leitbild für Grüne Gentechnik? in: Dirk Harreus (Hg.) Gentechnologie. Fakten und Meinungen zum Kernthema des 21. Jahrhunderts; Berlin 1999, 146, 195-210, 241f.

- (zusammen mit H. Schackert et al.) Molekularbiologie in der Viszeralchirurgie - prädiktive Diagnostik hereditärer Tumoren; in: Chir. Gastroenterologie 1999, 15, 195-201.

- Technische Weltanschauung bei Friedrich Dessauer; in: Johannes Rohbeck (Hg.) Philosophie und Weltanschauung, Dresden 1999, 140-155.

- Technische Entwicklung und sozialer Fortschritt; in: Renovatio 56 (2000), 99-106.

- Ethische Betrachtung der Anwendung bio- und gentechnologischer Verfahren in der Tierzucht; in: Deutsche Gesellschaft für Züchtungskunde e.V. (Hg.) Bio- und Gentechnologie in der Rinderzucht – Risiko oder Chance? DGfZ-Schriftenreihe 17; Bonn 2000, 56-66.

- Technological Development and social progress; in: Instituto del Filosofia Pontificia Universidad Catolica de Chile; Seminarios de Filosofia 12/13 (1999/2000), 41-52.

- Ethische Entscheidungsgrundlagen für die Energieentwicklung/-politik; in: Technische Mitteilungen. Organ des Hauses der Technik 3/01 94. Jahrgang, 133f.

- Gefangen in Sachzwängen? Zur ethischen Dimension der Gestaltbarkeit der Biotechnologie; in: St. Heiden et al. (Hg.): Biotechnologie als interdisziplinäre Herausforderung; Heidelberg/Berlin 2001, 83-96.

- La Teoria del Progresso nell'Illuminismo e la Rivoluzione Industriale/Fortschrittstheorie der Aufklärung und Industrielle Revolution; in:Studi Italo-Tedeschi/Deutsch-Italienische Studien; Meran 21 2000 (2002), 343-353.

- Künstliches Leben – Natur und technische Grenzen; in: W. Hogrebe (Hg.): Grenzen und Grenzüberschreitungen. 19. Deutscher Kongress für Philosophie; Bonn 2002, 865-872.

- Züchtung als technisches Handeln; in: A. Schäfer, M. Wimmer (Hg.) Machbarkeitsphantasien; Opladen 2003, 67-87.

- Künstliche Menschen? Posthumanität als Kennzeichen der hypermodernen Welt?; in: Ethica 11-2003-1, 3-32.

- La construccion gentecnica del hombre; in Ratio. Grupo de Investigacion Filosofica del Departamento de Filosofia de la UNMdP 2003; http://www.favanet.com.ar/ratio/actas.htm

- Gefangen in Sachzwängen? Zur ethischen Dimension der Gestaltbarkeit der Biotechnologie; in: F. Brickwedde et al. (Hg.): Biotechnologie – Innovationsmotor einer nachhaltigen Entwicklung; Berlin 2003, 119-137.

- Nachhaltigkeit als Ideologie?; in: Revista Portugesa de Filosofia 84/3 2003,763-784.

- Fortschritt und Risiko als Konstituentien technischen Handelns; in. J. Beaufort u.a. (Hg.) Fortschritt und Risiko. Zur Dialektik der Verantwortung in (post-)modernen Gesellschaften; FIF 21; Dettelbach 2003,53-83.

- Technologietransfer transkulturell als Bewegung technischer Kompetenz am Beispiel der spätmittelalterlichen Waffentechnologie; in: Wissenschaftliche Zeitschrift der Technischen Universität Dresden 52 (2003) Heft 5.6, 91-96.

- Epistemologie der Bio- und Gentechnologie; in: K- Kornwachs (Hg.) Technik – System – Verantwortung; Münster 2004, 285-297.

- Wie unnatürlich ist Doping? Anthropologisch-ethische Reflexionen zur Erlebnis- und Leistungssteigerung; in: C. Pawlenka (Hg.) Sportethik. Regeln, Fairness, Doping; Paderborn 2004, 279-291.

- Konzepte des impliziten Wissens und die Technikwissenschaften; in: G. Banse, G. Ropohl (Hg.):Wissenskonzepte für die Ingenieurpraxis. Technikwissenschaften zwischen Erkennen und Gestalten; VDI-Report 35; Düsseldorf 2004, 99-112.

- En camino nacia una metatecnologia; in: Erasmus. Revista para el dialogo intercultural: Tecnologia y Teologia; Ano VI-Nr.1-2004, 65-78.

- Der Cyborg als der Übermensch Friedrich Nietzsches? Anmerkungen zur Posthumanismusdiskussion; In. R. Kaufmann, H. Ebelt (Hg.) Scientia et Religio. Religionsphilosophische Orientierungen ; Fschr. für Hanna-Barbara Gerl-Falkovitz; Dresden 2005, 315-333.

- Innovationskulturen: Bedingungen technischer Kreativität; in: G. Abel (Hg.): Kreativität. 20. Deutscher Kongress für Philosophie 26.-30. September 2005 an der TU Berlin; Hamburg 2006, 290-300.
- Technik als Grundlage weltumspannender menschlicher Kultur. Technikphilosophische Reflexionen im Anschluss an Ortega y Gasset; in: Ch. Rodiek (Hg.) Ortega y la Cultura Europea; Frankfurt u.a. 2006, 101-118.

- Risiko – ein problemgeschichtlicher Abriss; in: Wissenschaftliche Zeitschrift der technischen Universität Dresden Bd. 55 (2006) Heft 3-4; Thema Risiko, 15-18.

- Über den Umgang mit technischen Risiken; in: Wissenschaftliche Zeitschrift der technischen Universität Dresden Bd. 55 (2006) Heft 3-4; Thema Risiko, 141-143.

- Ethical Acts (Actions) in Robotics; in: Ethics in science and Technology Vol 3 (October 2006) Hokkaido University (Sapporo, Japan), 50-66.

- Technology Transfer as Transcultural Modernization (Europe/South-East-Asia; in: Ethics in science and Technology Vol. 3 (October 2006) Hokkaido University (Sapporo, Japan), 67-79.

- Technik im Wertewandel. Wie entsteht Vertrauen in Technik? in: Forschung und Lehre 3/07 (14. Jg.), 140-142.

- Visions of Technology; in: Ubiquity 8, 10 March 2007, 1-6 (ISSN 1530-2180).

- Wegbereiter einer alternativen Moderne? Der Überwachungsstaat als Antwort auf Verunsicherung durch terroristische Umnutzung von Technologie; in: Ethica 15 – 2007 – 2, 145-172.

- Innovationskulturen, Technologietransfer und technische Modernisierung; in: Klaus Kornwachs (Hg.) Bedingungen und Triebkräfte technologischer Innovationen; Stuttgart 2007, 149-166.

- Technology Transfer and Modernization: What Can Philosophers of Technology Contribute?; in: Ubiquity 8, 48 (4.12.2007; ISSN 1530-2180), 1-15.

- Inferiority of Latin-American Technology? Consequences for Philosophy of Technology; in: N. Rehrmann, L. R. Sainz (HG.) Dos Culturas en Dialogo? Historia cultural de la naturaleza, la tecnica y las sciencias naturalesa en Espana y America Latina; Madrid, 2007, 249-268.

- Technik-Erwägungskultur und Leitbilder für die Grüne Gentechnologie; in: R. Busch u. G. Prütz (Hg.): Biotechnologie in gesellschaftlicher Deutung; Institut TTN; München 2007, 91-99.

- Medizin als Technoscience; in: Wissenschaftliche Zeitschrift der Technischen Universität Dresden 57 /2008, H 1-2, 17-19.

- Langzeitverantwortung und Nachhaltigkeit; in C. F. Gethmann, J. Mittelstraß (Hg.): Langzeitverantwortung. Ethik. Technik. Ökologie; Darmstadt 2008, 87-98.

VII. Systematic Treatise and Essays

- Im Anfang war der Egoismus: Die Soziobiologie als Neubegründung der Sozialphilosophie? in: A. Schöpf (Hg.): Aggression und Gewalt. Anthropologisch-sozialwissenschaftliche Beiträge; Würzburg 1985, 227-245.

- Biologie als Erste Philosophie? Überlegungen zur Voraussetzungsproblematik und zum Theoriestatus einer Evolutionären Erkenntnistheorie; in: Philosophische Rundschau 33 (1986), 103-121.

- Die Evolutionäre Erkenntnistheorie aus philosophischer Perspektive; in: August Fenk (Hg.); Evolution und Selbstbezug des Erkennens; Wien/Köln 1990, 83-106.

- Evolutionäre und philosophische Erkenntnistheorie. Kritische Bemerkungen zu reduktionistischen Deutungen des menschlichen Erkenntnisvermögens; in: Bernhard Dressler (Hg.); Der bedrohte Mensch. Zeitkritische Impulse christlicher Anthropologie, RPI Loccum, Arbeitshilfen Gymnasium 5 Rehberg-Loccum 1993, 42-52.

- Neuzeitliche Skepsis, nicht der Pyrrhonismus begründet Toleranz; in: Ethik und Sozialwissenschaften 5 (1994), 593-594.

- Selbstorganisation und Kognition: Fortführung oder Überwindung der Evolutionären Erkenntnistheorie? in: Werner Hahn, Peter Weibel (Hg.): Evolutionäre Symmetrietheorie - Selbstorganisation und dynamische Systeme; Stuttgart 1996, 193-202.

- Zur Aktualität der Kant'schen Metaethik in der neurophilosophischen Diskussion der menschlichen Freiheit; in: E. Graf, S. Hösel, S. Lingner, L. Leidl (Hg.) Wer braucht Kant heute?; Dresden 2007, 77-101.

VIII. Essays to the Historical Problems Investigations

- „Evolution" im 17. und 18. Jahrhundert - Fallstudien zur methodologischen Vorgeschichte von Darwins Theorie; in: Conceptus 42 (1983), 3-28.

- Zur Problemgeschichte des Topos „christliche Anthropozentrik" und seine Bedeutung für eine Umweltethik; in: Münchener Theologische Zeitschrift 37 (1986), 185-203.

- Metamorphosen eines „satirischen Realismus". Zweifelsexperimente zwischen Spätaufklärung und Frühromantik in Deutschland; in: Hans Körner, Constanze Peres, Reinhard Steiner, Ludwig Tavernier (Hg.); Die Trauben des Zeuxis. Formen künstlerischer Wirklichkeitsaneignung; Hildesheim 1990, 201-233.

- Renaissance-Philosophie als Wegbereiter neuzeitlicher Naturwissenschaft; in: Philosophischer Literaturanzeiger 45 (1992), 71-88.

- Aufklärungsphilosophie - Vielfalt der Ansätze und Methoden; in: Philosophischer Literaturanzeiger 49 (1996), 181-195.

- La Mettries Begründung der Anthropologie; in: Jan Beaufort, Peter Prechtl (Hg.) Rationalität und Prärationalität. Festschrift für Alfred Schöpf; Würzburg 1998, 81-92.

- Friedrich Nietzsche und die Zucht des Übermenschen (in bulgarischer Sprache); in Nietzsche im Osten; Vorträge anlässlich eines Kolloquiums der Philosophischen Fakultät der Universität Sofia und des Goethe Institutes am 27. November 2000; Sofia 2002, 89-104.

- Über den Philosophen Gustav Kafka; in: J. Rohbeck und H. U. Wöhler (Hg.) Auf dem Weg zur Universität. Kulturwissenschaften in Dresden 1871-1945; Dresden 2002, 139-151.

- Paracelsus Idee einer Bearbeitung der Natur. Anmerkungen zum Technikverständnis des Hohenheimers; In: Manuskripte, Thesen, Informationen hg. von der Deutschen Bombastus-Gesellschaft Nr. 22/2005, 13-18.

IX. Course Material

- Genolog 1999: Unterrichtsmaterial (E-Learning) der Überregionalen Frankfurter Sozialschule/Diözese Speyer; http://www.lernsystem.zsm.org

- Gentechnologie 1 – 12/2004 in Videolexikon: www.videolexikon.com

- Autonomie und Selbstbestimmung. Grundlegende Begriffe der Medizinethik und der Ärztlichen Praxis; Weiterbildender Studiengang Medizinethik der Fernuniversität Hagen 10/05, 53-94.

Dresden Philosophy of Technology Studies
Dresdner Studien zur Philosophie der Technologie

Edited by/Herausgegeben von Bernhard Irrgang

Vol./Bd. 1 Bernhard Irrgang: Technologietransfer transkulturell. Komparative Hermeneutik von Technik in Europa, Indien und China. 2006.

Vol./Bd. 2 Bernhard Irrgang / Sybille Winter (Hrsg.): Modernität und kulturelle Identität. Konkretisierungen transkultureller Technikhermeneutik im südlichen Lateinamerika. 2007.

Vol./Bd. 3 Lars Leidl / David Pinzer (Hrsg.): Technikhermeneutik. Technik zwischen Verstehen und Gestalten. 2010.

Vol./Bd. 4 Arun Kumar Tripathi (ed.): Bernhard Irrgang: Critics of Technological Lifeworld. Collection of Philosophical Essays. 2011.

www.peterlang.de

Zsuzsanna Kondor

Embedded Thinking
Multimedia and the New Rationality

Frankfurt am Main, Berlin, Bern, Bruxelles, New York, Oxford, Wien, 2008.
XI, 169 pp.
ISBN 978-3-631-57732-5 · pb. € 39.–*

The new devices of communication that have recently been emerging have far-reaching effects not only on our everyday lives, but also on our cognitive patterns: they lead us back again into the world of multimodality, and call attention, not incidentally, to the widening gap between everyday experience and the traditional convictions of philosophy. Traditional philosophical inquiries are seen in a new light when viewed from the perspective of communications technology. From that perspective, it becomes clear that a radical turn has become inevitable in the field of metaphysics and epistemology. This volume attempts to provide building-blocks for the new edifice of philosophy towards which that turn is leading.

Contents: Language and Communication · Philosophy and Communications Technology · Images and Words · Language and Cognition

Frankfurt am Main · Berlin · Bern · Bruxelles · New York · Oxford · Wien
Distribution: Verlag Peter Lang AG
Moosstr. 1, CH-2542 Pieterlen
Telefax 00 41 (0) 32 / 376 17 27

*The €-price includes German tax rate
Prices are subject to change without notice

Homepage http://www.peterlang.de